R. Kalley (Robert Kalley) Miller

The Romance of Astronomy

R. Kalley (Robert Kalley) Miller

The Romance of Astronomy

ISBN/EAN: 9783744763905

Printed in Europe, USA, Canada, Australia, Japan

Cover: Foto ©berggeist007 / pixelio.de

More available books at **www.hansebooks.com**

THE

ROMANCE OF ASTRONOMY.

BY

R. KALLEY MILLER, M.A.

FELLOW AND ASSISTANT-TUTOR OF ST. PETER'S COLLEGE, CAMBRIDGE.

London:
MACMILLAN AND CO.
1873.

PREFACE.

THE greater part of the following papers was originally written for delivery in the form of popular lectures. They were then published in a University Magazine, the *Light Blue;* and having met with considerable success in both these ways, they have now, at the suggestion of several friends, scientific and non-scientific, been partially re-written and enlarged into their present form.

I have endeavoured, however, to keep their original object unaltered, and to write nothing which would not be at once interesting and intelligible to non-scientific readers. There is no lack of systematic, and yet easy, works on Astronomy, such as those of Sir John Herschel, M. Arago, and Mr. Norman Lockyer; and I have, therefore, made it my object not so much to instruct as to entertain, and possibly in some cases to inspire a taste which might lead to the further prosecution of a most fascinating study. This must be my apology for passing over entirely many important parts of the subject, and simply selecting a few points here and there which seem to afford scope for striking or amusing amplification.

Since the following sheets have passed through the press, I have learnt that Professor Adams and others have thrown grave doubts upon the accuracy of the calculations upon which Professor Hansen's theory of eccentric gravitation at the moon was founded. Should this theory fall to the ground, the argument for the habitability of our satellite which was founded upon it, and which I have explained at page 65, must go with it. But it will remain as a striking and interesting episode in the history of scientific speculation.

In the note to page 53 the fact of the moon's always turning the same face towards us is spoken of as a question of Rigid Dynamics. But it is possible that it may rather depend upon the earth's action on the moon while in a viscous state.

I have to thank two very distinguished Members of my own College, Sir William Thomson and Professor Tait, for kind suggestions and advice.

PETERHOUSE,
December, 1872.

CONTENTS.

	Page
INTRODUCTION	1
THE PLANETS	5
ASTROLOGY	40
THE MOON	50
THE SUN	71
THE COMETS	85
LAPLACE'S NEBULAR HYPOTHESIS	96
THE STARS	112
THE NEBULÆ	135

ERRATA.

Page 7, line 20, *for* heavy, *read* massive.
Page 90, line 21, *after* them, *insert* nearly.
Page 130, line 11, *for* million, *read* thousand.

The Romance of Astronomy.

THE Romance of Astronomy strikes one at first as sounding something very like a contradiction in terms. We might naturally be inclined to think that there is about as much of romance in astronomy as there is of poetic fire in Martin Tupper, or of charity in a Saturday Reviewer. Any one listening to the conversation of two astronomers, and hearing them descanting enthusiastically about perigees, apogees, and syzygies, right ascensions and declinations, precession of the equinoxes, and the longitude of the moon's ascending node; or any one opening at random the pages of a work on the science, and finding an incomprehensible mass of calculations, formulæ extending over twenty lines and using up all the letters of two or three alphabets, and diagrams like nothing in the earth beneath, or in the waters under the earth, and only bearing a very faint resemblance to things in heaven above; any one we repeat, on getting such an introduction to the subject, would be very much tempted to think that romance and astronomy were altogether incompatible. Science is said by rhetoricians to be the logical opposite of poetry, and whence then can come any element of romance into the sternest and loftiest of the sciences?

But if we consider not so much the study of the science itself, in its profound and recondite details, as the results to which it attains, the magnitude and importance of the subjects it treats of, and the beauty and grandeur of the phenomena it investigates, we shall have to acknowledge that somewhere or other in the ponderous tomes of astronomical science there must lie entombed rich stores of novel and unwonted interest. The science which fathoms the infinite and reckons up the eternal, which pierces the abysses of space, grasps the orb which we see now by the light that left it eighty thousand years ago, measures its distance, and traces its movements—the science which accomplishes such marvels as these, and the history of the great men who achieved these noblest triumphs of human intellect—must surely furnish many themes and contain many episodes of a character as wonderful and as truly romantic as we can find within the airy realms of fiction or of poetry. And besides the grandeur of the phenomena of astronomy and the romance which gathers round its history in all ages and casts a brilliant gleam here and there upon its sober annals, there often flashes even across the pages of the driest and most mathematical parts of the subject a glimpse of strange and unexpected interest; and a fact here and a figure there will start the mind in a train of fresh and novel speculation, and set the fancy to luxuriate in new and untrodden realms. Many of these points moreover to which we allude, though very interesting and wonderful in themselves, are yet of comparatively little importance from an astronomical point of view; their interest centres in themselves, and the results to which they lead must be regarded as rather curious than valuable; and hence they are but little to be met with in books, or if touched upon at all, are soon abandoned with the remark that it is time to quit such regions of endless and unavailing speculation. Now some of these speculations we purpose following out a little to their legitimate conclusions, trusting that from the above reason they may prove new to many of our readers. And in the other points which we take up—for we must not confine ourselves to so limited a portion of the romance of astronomy as this alone—we shall seek to select those which are likely to prove at once the most striking and the least familiar to non-scientific readers.

THE PLANETS.

We turn naturally first to our sister planets. They are in all respects analogous to our own globe; they hold the same position in the great system of the universe that we do, and in them—if in any of the orbs of heaven at all—we might expect to find the face of nature presenting the same appearance, and the course of nature the same phenomena, that they do to us. But not such do we find to be the case. Some of them indeed will resemble us pretty closely in one thing and some in another, but in every one the points of contrast will be much more numerous and striking than those of similarity.

In looking over a table of the elements of the planets, one of the points which most attract our attention is the very great differences in size which they present; and as this circumstance is the cause of some of their most striking physical pecularities, we may commence with it our examination of them.

It affords, too, a remarkable illustration of the statement we have made, that a fact of apparently little importance in itself, often leads indirectly to very unexpected and startling consequences. The magnitude of a planet is a point we should never expect to find in any way necessarily connected with the nature of the beings who inhabit it and the general character of life at its surface, and yet we shall find it intimately related to these matters, and that to the production of very singular consequences indeed. Take for instance the case of one of the minor planets—Ceres, or Pallas, or Vesta. Astronomers tell us that the diameter of the earth is 7912 miles, and that of Ceres 160 miles; and the words may very easily pass in at the one ear and out at the other, without leaving any impression behind; or if we pause for a moment to think over them, it will likely only occur to us what a compact little world Ceres must be, how easy it must be to get from one place to another in it, and how delightful to be able to sail round the world, pay a visit to one's friends at the Antipodes, and get settled at home again,—all within the short space of a week. But if we look at the subject a little more closely, we shall find that it involves far more extraordinary consequences than these. We know that by the law of gravitation, the force with which one body attracts another varies directly as its mass, and inversely as the square of its distance; and also that a sphere attracts any external object as if its own mass were all collected at its centre. Now the diameter of the earth being fifty times as great as that of Ceres, it is altogether 125,000 times as large; but this disproportion being partially counteracted by the greater distance of its surface from the centre, it follows that on the whole the force of gravity here is fifty times greater than at Ceres—or, in other words, any object here is fifty times as heavy as it would be there. Now let us look for a moment at what is implied in this. The first and most obvious consequence is, that a man will be able to lift fifty times as great a weight there as here. A ton would be an easy load, boys would play at ring-taw with huge round boulders instead of marbles, and a rattle intended for a stout baby might be made as heavy as a moderate sized cannon-ball. If the tower of Siloam had fallen there instead of here, the men, instead of being crushed by its weight, would have lifted themselves and it up with the greatest ease, and felt nothing the worse for the accident. But there are more singular consequences yet. We know that if a body be once set in motion, it would continue moving to all eternity, if not brought to rest by some external force. Thus when a man leaps up into the air, he would continue ascending for ever, were it not for the attraction of the earth, which very speedily brings him down again. But at Ceres, this force is so small that it will be much longer before it takes effect, and a man might consequently leap to an enormous height before the attraction would check his ascent. Jumping over a house-top would be a very trifling exploit, while a good leaper would think nothing of clearing, with a short run, the new tower of St. John's Chapel, or the Great Pyramid itself. Staircases might be abolished, for even a stout old lady could easily jump in at a three-story window. The range of projectiles would be increased in proportion. Ensign Humphry, with a good telescope, would put a ball into the bull's-eye from a distance of twenty miles. An economical war-minister could no longer build on the security afforded by "the streak of silver sea," for Great Britain might be swept with artillery from the Land's End to John O'Groat's House, by batteries erected far inland on the continent.

Nor have we exhausted the wonders of Ceres yet. When Swift made Gulliver describe his adventures among the Brobdingnagians, he probably had no idea that they were even farther removed from reality than the other creations of his fancy—that they were not only myths, but absolute impossibilities. A giant here would be crushed by his own weight. A very easy calculation will show this. Suppose a man twelve feet high, and stout in proportion. He will be twice as long, twice as broad, and twice as thick as an ordinary mortal, and thus eight times as heavy. Now if we take a cross section of his leg, the cut surface will be twice as broad and twice as wide as usual, and thus four times as large altogether. We shall thus have eight times the ordinary weight to be supported by only four times the ordinary surface; and hence the stress on the bone will be twice as intense as usual. In the same way, in a man three times the ordinary height, the stress would be three times as great, and so on. Such a stress might perhaps be borne, but when we got the length of a giant sixty feet high, the stress would be ten times as great, and that the bone certainly could not bear. It would either be crushed outright if the giant attempted to stand erect, or else his legs would totter, his knees would bend, and his mighty body come thundering down to the ground. Once down, it would be utterly impossible for him to get up. A sitting posture he might perhaps compass; but if he were a very big giant indeed, that too would be out of the question—and he could do nothing but lie along on the ground. But transport him to our queer little friend Ceres, and he is all right at once. In a moment he becomes fifty times lighter than he was, he leaps to his feet with ease and rears his huge head sixty feet into the air, his legs recover their strength, his aching bones grow well, and he may proceed, if he please, to astonish the acrobatic natives of the planet by gymnastic exploits far surpassing even their own.

Indeed, all the wonderful feats we have seen that an ordinary man would be capable of at the surface of Ceres, must be multiplied fifty-fold when we take into account the superior possible size of the inhabitants of that planet. Muscular exertion there goes fifty times as far as it does here; and as these gigantic beings will be able to put forth at least fifty times as much of it, the exploits they will be capable of achieving must be no less than 2500 times as great as anything that could be done here. Upon this enlarged field of speculation we can scarcely venture to enter. The wildest flights of fancy, and the most exaggerated visions of fairyland, will be more than realized. Like Milton's angels, they could tear up the hills by their bases, and hurl them at their foes. Stronger than the vanquished Titans of old, fetters of iron would be to them as threads of gossamer; and mountains piled on the top of mountains would not suffice to crush or imprison them with their load. Like the genii of the Arabian Nights, they could spring at a bound from the earth to the clouds, or clear half-a-dozen miles at a single leap. The seven-league boots would be no longer a fable. Puck said he would put a girdle round the earth in forty minutes; but one of these giants of Ceres would stride round his planet in less than half the time.

Of course all the other denizens of the asteroid will have their size and strength increased in the same proportion. The racehorse will rear his crest two hundred feet into the air, and gallop five thousand miles an hour. The giraffe on the plain will lift his stately head, and browse on the trees that crown the mountain-top. The ponderous elephant will cover three acres of ground, and surpass in strength the most powerful steam-engine. The lion's roar will be more dreadful than the thunder-peal, and his resistless spring more terrible than the lightning's flash. Snakes two hundred feet in circumference and a thousand in length, will roll their huge coils through the forests; while the sea will boil and foam with the gambols of its mighty inmates, and the gigantic carcase of Leviathan extend for a mile along the deep.

If we reverse the circumstances and go to a world larger than our own instead of smaller, the case will of course be exactly the opposite. If we ourselves were transported to the sun, we should feel as much like fish out of water as the colossal inhabitants of Ceres would do here; and in fact it will be readily seen that if the sun were inhabited by beings constituted like ourselves, its population could consist only of dwarfs two or three inches in height. Very singular it surely is that the larger the world, the smaller its denizens must be, that the inhabitants of the earth should be men, those of the sun dwarfs, and those of the tiny asteroid giants.

We must remind our readers—what they might well be excused for forgetting—that we are not romancing about what might be the case in some absurd and impossible circumstances, and if the laws of nature were to undergo some extraordinary and unheard of change, but that we are speaking in all truth and soberness, and that what we have stated is absolute and demonstrable fact.* If any man were transported at this moment to the planet Ceres, he would be able to do everything we have mentioned; and the actual inhabitants of that planet, if constituted like ourselves, must be able to do the same. Whether, if they exist at all, they are beings like ourselves or not, of course we cannot tell; their frames may be feebler and their powers more limited than our own, and life at the Asteroids may be after all not so very different from life on the earth itself.

And now to consider a few other points connected with the planets—those namely which arise from their various positions relatively to the sun, and from the character and velocity of their movements. The general celestial phenomena, and the periodical changes connected with them, must of course be the same at all the planets. They have the same alternation of day and night, of summer and winter, that we have. For them, as for us, the sun has been set to rule the day, and moons and stars to rule the night. But though their times

---

* See Herschel's Astronomy, end of Chap. VIII, where some of the above ideas are hinted at. Our mathematical readers will see that there is not the slightest exaggeration in the extent to which we have carried them.

and seasons, their days and years, are exactly analogous to our own, yet the differences in their positions and movements will produce corresponding differences of a very marked kind in the lengths of those periods and in the vicissitudes of climate occasioned by them. The most important of these differences are caused of course by the very various distances from the sun at which the planets are situated. Mercury is three times nearer it than we are, and Neptune thirty times farther away. It follows from this that at Mercury the sun will appear nine times as large as it does to us—the intensity of its light and heat being of course increased in the same proportion; while at Neptune all its influences will be nine hundred times feebler than they are here. Hence at the former planet the average heat must be greater than that of boiling water; and if at its creation it contained any seas or rivers like our own, they must have been long ago dissipated in vapour by the sun's overpowering beams. At Neptune, on the other hand, that luminary will appear no larger than one of the planets does to us. How cold and drear an abode it must therefore be!—its brightest noonday more dusky than our winter twilight, and its hottest midsummer far colder than our frozen poles.

Another consequence of the varying distances of the planets is a great diversity in the length of their years, some of them being as short as three of our months, while one extends over no less than a hundred and sixty years. How long and dreary the circle of the seasons must be there!—forty years of spring, forty of summer, forty of autumn, and forty of winter. The contrast between the seasons will be in some of the planets greater, and in some much less than our own; at Jupiter especially there will be no perceptible change of seasons at all, and day and night will everywhere last for twelve hours each, just as at our equator. The orbit of Mercury presents a very marked eccentricity;—in other words the planet is much nearer the sun at one period of its revolution than at another; so much so that that luminary will appear twice as large, twice as bright, and twice as hot, when Mercury is in perihelion as when in aphelion; a circumstance which cannot fail to be productive of very serious effects to its inhabitants. Even at our own earth, whose orbit is so much more nearly circular, the same cause produces a quite perceptible effect. The earth is nearest the sun in December, and the consequence of this is that in our northern hemisphere the winter is rendered milder than it would otherwise be, while south of the equator the heat is considerably aggravated. In June the opposite will be the case, and the whole result is evidently to make the northern hemisphere more temperate than the southern. Accordingly we find that the intense heat of the sun is much more complained of in the Australian and South African deserts than in those to the north of the equator. The eccentricity of the earth's orbit is at present diminishing at a small uniform rate,* and the effect of this, in a sufficiently long course of time, would be to decrease these annual variations of temperature. In some of the other planets, however, it is on the increase, and when this fact was first discovered, it excited great interest among astronomers. The increment, though extremely small, appeared to be perfectly regular, and if continued long enough it must infallibly cause such frightful vicissitudes of cold and heat as to destroy any life which might exist at their surfaces. Lagrange, however, succeeded in establishing a beautiful and simple relation between the eccentricities of the planetary orbits, which showed that none of them could ever exceed certain definite limits, and that although they might increase

---

* Due to the perturbing influence of the other planets.

for almost countless ages, a maximum would in time be reached, and a compensating period of diminution would ensue.

Lastly, the rotations of some of the planets on their own axes are performed in much shorter periods than that of the earth. The effect will be to shorten the length of the day, to make the planet bulge out at the equator, and to diminish gravity by reason of centrifugal force. We all know that if a stone be tied to a string and whirled round, it will acquire a tendency to fly off, which will be greater the faster it is whirled. In the same way some of the planets spin round so rapidly as to communicate to any body on their surfaces a very powerful tendency to fly off, which is however counterbalanced by the effect of gravity. But if Jupiter's rotation were only four times faster than it is, the centrifugal force would be so great that all the inhabitants would be sent flying off through the air—or rather along with it, for it would go too. When the impulse with which they started was lost, they would of course fall back to the ground, but only to be shot off again at once; and in this state of perpetual oscillation, bouncing up and down like an india-rubber ball, they would spend all their lives, unless they took some means of anchoring themselves to the surface of their planet.

The class of phenomena which we have been last considering depend all of them upon the positions and movements of the planets, and are hence common, with various modifications, to the whole of them. But besides these there are connected with all of them special points of individual interest, arising from circumstances peculiar to themselves alone, and over these we must cast a rapid glance before we proceed in our excursion to visit a new set of worlds.

Of the first of the planets, Mercury, we know but little. From the closeness of his proximity to the sun he can never be seen with the naked eye, except occasionally for a few moments close to the horizon, immediately after sunset or before sunrise; and even these hurried glimpses cannot be got except at considerable intervals and under very favourable circumstances.* Hence, though his existence seems to have been known from a very early period, he was comparatively seldom seen before the invention of the telescope. Copernicus lamented upon his death-bed that he had never been able to catch a glimpse of Mercury at all; the mists from the marshes of the Vistula too obstinately fringed the morning and evening

---

* It is calculated that Mercury, Venus, and the Earth will, from a similar reason, never be visible at all from the surface of Uranus.

horizon round the Observatory of Thorn. A distinguished French astronomer of the same period only saw him twice. The telescope when turned upon him shows us little but a small round disc, which exhibits phases, like the moon, according to its relative positions with regard to the sun and to the earth. Recent observations have revealed enormously lofty mountains upon his surface, eight times as high in proportion to their planet as the Himalayas are to our own globe. The proximity of Mercury to the sun, the eccentricity of his orbit, and the fact that he is unattended by any satellites, rendered the determination of his mass and other elements a matter of much difficulty, and great discrepancies exist between the earlier estimates of them. Fortunately his small size, and the consequent insignificance of the perturbations he produces in the other planets, diminished the importance of having an accurate knowledge of him. Any similar uncertainty about one of the larger planets would have interposed most serious obstacles to the progress of science, and would, for example, have rendered the discovery of Neptune impossible.

It is at present uncertain whether there are any planets within the orbit of Mercury. If there are, their light must be so overpowered by that of the sun, as to render them visible only when he is under eclipse, or when they are passing across his disc, in which case they would appear as small black spots. Astronomers have occasionally fancied that they detected planets under the latter circumstances, but they have never felt certain that what they saw were not merely some of the ordinary spots on the sun. A French astronomer, M. Lescarbault, felt pretty confident on one occasion that he had found a real planet, to which he gave the name of Vulcan; but twenty years have passed away, and the discovery has never been confirmed. It was hoped that at the recent total eclipse Vulcan might have been seen near the edge of the moon's disc when the sun's light was cut off; but if he really exists, he lost the glorious chance then offered him of proving the fact, by perversely hiding behind the sun, or between it and the moon.

With Venus we are all familiar. It is the most brilliant of all the planetary or stellar orbs; and the "Star of the Evening, Beautiful Star," has been sung by poets of every age and clime, from Homer to the Christy Minstrels. Like Mercury, and for the same reason, Venus is seldom seen except about sunrise or sunset; but as her elongation from the sun, though limited, is much greater than that of Mercury, she is very frequently visible. Sometimes even, though at rare intervals, she is sufficiently near us to be seen when the sun is above the horizon; and the sight of the little planet, shining softly out in fearless companionship with the dazzling orb of day, is described as singularly striking and beautiful. Varro relates a tradition that Venus shone thus at noonday, a most auspicious portent, upon Æneas' voyage from Troy to Italy. And on the occasion of one of the first Napoleon's triumphal entries into Paris after a successful campaign, Venus joined in the pageant of the procession; exciting the intensest enthusiasm among the populace, who regarded her daylight appearance as a miracle; and flattering even the stern heart of the conqueror with the thought that Heaven itself had sent its fairest orb to grace the brilliance of his triumph. It was long before it was discovered that the morning and the evening star were one and the same planet, and hence we meet with it in the classics under a double name,—Lucifer, Son of the Morning, and Hesperus, Star of the Eve.* A similar confusion prevailed with

---

* Ἠμος δ' Εωσφόρος εἶσι, φόως ἐρέων ἐπὶ γαῖαν,
Ὀυ τε μέτα κροκόπεπλος ὑπεὶρ ἅλα κίδναται Ἠώς.
      Homer Il. 23. 226.
Ἑσπερος ὃς κάλλιστος ἐν οὐρανῷ ἵσταται ἀστήρ.
      Homer Il. 22. 318.

regard to Mercury, which as a morning star was styled Apollo, the Lord of Day, and as an evening star Mercury, the Patron of Robbers.

The phases of Venus are readily shown by the telescope, and were detected by Galileo soon after the invention of that instrument. Delighted at his discovery, but unwilling to publish it until verified by fuller observations, he shrouded it in the following line:

Hæc immatura a me jam frustra leguntur,*
which, anagrammatically transposed with a little license, gives

Cynthiae figuras emulatur mater amorum.†
This ingenious way of embalming a discovery until ripe for publication was a favourite one with the mediæval astronomers, as it enabled them to claim priority, if anyone else, by making the same discovery, should take the wind out of their sails.‡ The result of Galileo's first observation upon Saturn was communicated to the scientific world in the form

aaaaabeeegiiiillmmmmmnnoprrstttuvv,

---

* These things, yet unripe and not understood, are read by me.
† The mother of loves emulates the phases of the moon.
‡ Simon Mayer, a Bavarian astronomer, contested with Galileo the priority of discovery of Jupiter's satellites, but his claim appears to have been not only unfounded but absolutely dishonest.

letters, which he afterwards arranged thus :—

Ultimam planetam trigeminam observavi.*

Huyghens' discovery of the real nature of the ring was first made known thus:

aaaaaaa ccccc d eeeee g h iiiiiii llll mm nnnnnnnnn oooo pp q rr s ttttt uuuuu,

which, when he had fully satisfied himself of its truth, he interpreted into

Annulo cingitur tenui plano nusquam cohœrente ad eclipticam inclinato.†

The dazzling and uniform brilliancy of the disc of Venus, which renders it very difficult to get a good telescopic view of it, is supposed to be caused by the reflection of the sun's rays off a dense cloudy stratum; and in fact it seems probable that we never see its surface at all, but only its illuminated atmosphere. In Mars, on the other hand, which is the next planet, we can trace with perfect distinctness the outlines of continents and seas. The bright ruddy light which distinguishes this planet from all the others proceeds from its solid parts, and is caused doubtless by a prevailing reddish tinge in the soil, something the colour of our red sandstone, only much

---

* I have perceived the most distant planet to be threefold.

† It is surrounded with a thin plane ring, nowhere adhering to it, and inclined to the ecliptic.

brighter. The seas are distinguished by their blueish tinge, while at the north and south poles are large and irregular patches of a brilliant white. These have been conjectured with great probability to be vast tracts of ice and snow; and this idea is confirmed by the fact that they are of variable size, being largest during their winter, and diminishing very perceptibly on the approach of summer.

Leaving this planet of the "Red, White, and Blue," and passing over the asteroids, to which we shall return presently, we come to Jupiter, the largest and most important of all the planets. This great orb is no less than thirteen hundred times as large as our earth, and everything connected with him is on the grandest scale. His years last for ten thousand days, his motion on his axis is so rapid that the heavenly bodies must be seen changing their places every minute, and his nocturnal sky is illuminated by a band of four large and beautiful satellites.* His surface is divided into bright and dark belts

---

* These satellites have played a very important part in the history of science. Their discovery was hailed as a valuable confirmation of the Copernican theory of the solar system, of which they present a miniature picture. They have proved of great service to the navigator; the time of their eclipses can be calculated with great accuracy, and, when compared with local time, gives a simple method of determining that important

parallel to the equator. The former are supposed to represent dense masses of clouds, reflecting the sun's rays more perfectly than the solid body of the planet. Their parallelism to the equator, and their comparatively uniform breadths, are probably to be accounted for by steady atmospheric currents, of a character similar to our trade and return trade winds, but much more violent, in consequence of Jupiter's more rapid rotation on his axis. In fact all the observations upon his atmosphere tend to show that the wind blows at his surface with overwhelming fury, sometimes surpassing a thousand-fold our most terrific hurricanes.

The moons of Jupiter were among the earliest revelations of the telescope. They were discovered by Galileo, who at first supposed them to be stars, and was much puzzled for a few nights by the irregular manner in which Jupiter appeared to move about among them. He had great difficulty in getting the scientific world to acknowledge their existence. Some of the contemporary philosophers thought that they were optical illusions due to an imperfection of the instrument. Many absolutely

---

and difficult geographical element, the longitude. And some discrepancies between their calculated and observed positions first suggested the great discovery of the finite velocity of light.

refused to look through such an unnatural and diabolical engine as the telescope, and of course there was no other way of proving to them that the moons were really there.* One of these sceptics, Libri of Pisa, died during the heat of the controversy; and we find Galileo, in a letter to a friend, charitably hoping that the way to heaven lay past the planet Jupiter, and that Libri might be convinced at last. Another unbeliever, a rather eminent astronomer of the name of Sizzi, delivered an elaborate harangue against Galileo, which is still extant, and in which he argues as follows:—"There are seven windows given to animals in the domicile of the head, through which the air is admitted to the tabernacle of the body, to enlighten, to warm, and to nourish it;

---

* It is not quite certain that Jupiter's satellites have not occasionally been seen with the naked eye by persons of very powerful sight. In an early Japanese plate Jupiter is represented with two small stars beside him, which very possibly are meant for two of his moons. At a time when this subject happened to be exciting a little discussion in the scientific world, a German lady declared that she could see one of the satellites. Unfortunately for her probity, it was soon found that she always saw it on the wrong side of the planet—to the right when it should have been to the left, and *vice versâ*. The explanation was easy. She had got hold of some diagrams representing the apparent relative positions of Jupiter and his satellites from day to day, but they were constructed for using with the common astronomical telescope, which is an inverting one.

which windows are the principal parts of the microcosm, or little world—two nostrils, two eyes, two ears, and one mouth. So in the heavens, as in a macrocosm, or great world, there are two favourable stars, Jupiter and Venus; two unpropitious, Mars and Saturn; two luminaries, the Sun and Moon; and Mercury alone, undecided and indifferent. From these, and from many other phenomena of nature, which it were tedious to enumerate, we gather that the number of planets is necessarily seven. Moreover, the satellites are invisible to the naked eye; and therefore can exercise no influence over the earth; and therefore would be useless; and therefore do not exist. Besides, as well the ancient Jews and other nations as modern Europeans, have adopted the division of the week into seven days, and have named them from the seven planets. Now if we increase the number of planets, this whole system falls to the ground."

Absurd as this tirade is, we wonder at it the less when we find the illustrious Huyghens talking in a similar strain after his discovery of the first satellite of Saturn. He says:—"The solar system is now complete. It consists of six planets and six moons, and from this equality, and from the fact that they together constitute the perfect number twelve, we infer that no more satellites will be discovered."

The philosophers both of the ancient and middle ages had great belief in perfect numbers, but their superstitions have, in the nineteenth century, been thrown completely into the shade by the wild ravings of Comte, the high priest of Positivism, about primes. Like Sizzi, he had a great partiality to the number seven, because it was a prime, and because it was "composed of two progressions followed by a synthesis, or of one progression between two couples." For these reasons he wished it to be made the basis of our scale of notation. The latter reason we frankly confess our inability to comprehend; the former is intelligible, but singularly inconsequential. Most people would think a prime the worst number possible to found a scale on. His favourite number of all, however, is thirteen, and that for the following reasons: It is a prime; it is the seventh prime; seven is a prime; it is the fifth prime; and five is a prime. Here unfortunately he has to stop; five is the fourth prime, and four, on Comte's principles, is a very poor number indeed. It is a perfect square, and nothing on earth can twist that into a prime. Comte sincerely regrets this little flaw; if only twice two did not make four, thirteen would be an absolutely perfect number. Still it is so near it that it cannot be so very unlucky as it is popularly considered; and we trust none of our readers will ever again think it necessary to count the number of guests at a dinner-table.

Undeterred by the cogent arguments of Sizzi, Galileo, so far from giving up his moons or abandoning his infernal machine, turned his telescope, after investigating the orbits of Jupiter's satellites, to other bodies of the system, and soon detected those most extraordinary appendages of the next planet, the rings of Saturn. The highest magnifying powers show these rings merely as thin luminous threads crossing the disc of the planet and projecting slightly beyond it at either side, but to the inhabitants of Saturn itself their appearance must be inconceiveably grand. To the dwellers on one side of the planet the rings must present the magnificent spectacle of two vast luminous arches spanning the sky from horizon to horizon and rotating with enormous velocity;* and to the people on the other side the appearance will be the same, only that the arches will be dark instead of bright; while the regions which lie beneath their shadow will be plunged for fifteen years at a time in perpetual night. The feeble telescope with which Galileo discovered

---

* If it were not for this rotation they could not remain in equilibrium, but would be precipitated upon the surface of the planet.

the rings only revealed to him two protuberances beyond the disc of the planet at the opposite ends of a diameter. They appeared to him to be detached bodies, and he was much surprised to find that they did not change their positions relatively to the planet, and therefore neither revolved round it nor rotated with it in its daily course. But extraordinary as this phenomenon appeared, it became still more so when these two objects gradually diminished in size, and finally disappeared altogether. Galileo was utterly baffled. "Is the legend of mythology," he asked in amazement, "no longer a fable, and has Saturn really devoured his children?" The explanation of course was that the planet, advancing in its course, and changing its position relatively to the earth, had brought its equator into the same plane with us, so that the rings only presented their narrow rim to us, instead of their broad flat surface. But it was not till long afterwards that Huyghens, with improved telescopes, detected their real nature. Maupertius started a quaint theory for their origin. He supposed that they might be the mangled remains of an unfortunate comet, which had incautiously come too near Saturn, and got his tail wound round the planet and twisted off. A more probable theory we shall meet with further on.

Till within a comparatively recent period these five planets, Mercury, Venus, Mars, Jupiter, and Saturn, were believed to be the only ones besides our own earth in the system, but in the year 1781 Uranus was added to the number by Sir William Herschel. He did not suspect at first that it was anything but a comet, but as every observatory in Europe immediately set to work to calculate its orbit, it was soon recognised as a planet. Herschel wished to call it Georgium Sidus, after his kindly and munificent patron, George the Third. Several of his brother astronomers urged that it should be named after the illustrious discoverer himself, but the advocates of uniformity insisted upon the classical nomenclature being adhered to. The rival claims of all the old gods and goddesses were discussed. The name of Neptune found considerable favour in this country, Englishmen being then justly proud of the exploits of their fleet,* but the foreign astronomers would not agree to this. Many other names were suggested, and backed up by fanciful and epigrammatical reasons. Uranus was finally adopted, on the suggestion of Bode that the most distant of the

---

* Would the present Admiralty like to have a newly-discovered planet christened Megæra?

planets might appropriately be called after the most ancient of the gods.

It was soon found that the planet had been observed no less than nineteen times before in different parts of the heavens, but from its great distance, and consequently insignificant apparent magnitude, it had always been mistaken for a star.* This remarkable discovery excited the greatest interest among astronomers, and the hope began to be entertained that other distant planets also might have been mistaken for stars, and that the number of the planets might be thus still further added to. The only other discovery however which has yet been made of the character anticipated is that of Neptune, whose existence was first suspected by Bouvard in 1821, from the perturbations in the motions of Uranus caused by his disturbing influence. The problem of determining from these scanty data the distance, the orbit, and the mass of the disturbing planet, was evidently a possible one; but the analytical difficulties which it presented to the mathematician were so

---

* Lemonnier, in especial, seems to have narrowly escaped detecting its real nature, as he had observed it several times. But his observations were not registered and compared with sufficient care to lead to any results; indeed one of the most important of them was afterwards found by Bouvard scribbled upon a confectioner's paper bag.

enormous, that for more than twenty years no one attempted to grapple with them. Our own University had the great honour of first undertaking the task, and of prosecuting it to a successful conclusion. Mr. Adams commenced his ever-memorable researches immediately after taking his degree in 1843, and on the last of September, 1845, his calculations of the place in which the supposed planet should be sought for were tendered at Greenwich Observatory. Before commencing the search, which was likely to prove a laborious one, the Astronomer Royal requested Mr. Adams to make some further calculations, with a view of confirming his results;* but while he was engaged on these, M. Le Verrier (who had been, unknown to both of them, employed in similar researches) published the results of his calculations on the first of June, 1846. As they agreed exactly with Mr. Adams', Professor Airy's hesitations were removed, and he wrote to Mr. Challis, recommending a careful search with the great Northumberland refractor in the Cambridge Observatory.

---

* Mr. Adams had based his calculations on the perturbations of Uranus in longitude, and Professor Airy suggested that he should examine whether those in radius vector would lead to the same results.

D

This advice was immediately followed, and an accurate map of the part of the heavens in question was commenced, with the hope that on a second survey, some star in it would be found to have changed its place, and thereby shown itself to be the planet sought for. But before this labour was completed, Dr. Galle, a Prussian astronomer, who had the advantage of having a good map already in his possession, found a new star not laid down in his chart; and a little investigation established this at once as the long sought for orb. Professor Challis found that it was one of the bodies he had already mapped down, and that a few nights more must have infallibly led to its discovery by him also. Considerable jealousy was felt at the time between England and France with regard to the priority of claim between Adams and Le Verrier, the French astronomer being much disappointed to find that our countryman had vanquished the difficulty first, although his discovery was not made public at the time. But after all, the question of priority is a small one; each of the astronomers completed the task by his own unaided genius, and the names of Adams and Le Verrier will be handed down to posterity with equal honour, as the solvers of the hardest mathematical problem which has yet engaged the attention of scientific men.*

No planet more distant than Neptune has yet been discovered; but about sixty tiny orbs have been added to the system, whose existence had been previously unsuspected—not from their distance, but from their minuteness. We allude of course to the asteroids. The history of their discovery is very interesting, and affords a remarkable contrast to that of Neptune; being the result of a bold and fortunate guess, while the other was the fruit of years of

---

* The problem was the solution of a series of simultaneous partial differential equations with nine unknown quantities, namely the mass, mean distance, eccentricity, epoch, and perihelion longitude of the unknown planet, and the corrections to the latter four elements of Uranus. The smallness of the perturbations in latitude showed that the inclinations and nodes might be neglected, or, otherwise, the number of unknown quantities would have been thirteen. Many of our readers will understand the impossibility of solving such a problem by any ordinary mathematical methods, and even the usual devices of the Planetary Theory, evolved by the genius of Laplace and Lagrange, failed in application in consequence of the inverse character of the problem. In fact, the old armoury of Science was unavailable, and Adams and Le Verrier, in fighting their great battle with Nature, had to invent a fresh weapon for every stage of the conflict. For an interesting sketch of their labours we may refer our mathematical readers to Grant's "History of Astronomy," while the question of priority will be found discussed in Airy's "Historical statement of circumstances connected with the discovery of the planet beyond Uranus."

patient toil. Soon after the elements of the planets came to be accurately known, a remarkable empirical law was observed to connect their several distances from the sun. These were found to form a series, the difference between each of whose terms was twice as great as the preceding difference; in other words, the distance of any planet from the next without it was twice as great as its distance from the next within it. The only exception to this rule was in the case of Mars and Jupiter, whose distance from each other was much too great; in fact, it seemed as if there was a planet wanting between them to complete the perfect series. This fact, which was first noticed by the Baron de Zach, was considered so remarkable, that a company of astronomers banded themselves together to institute a search for the missing orb, and shared out among themselves the part of the heavens in which it was expected to be found. The leading men of the day considered the idea as altogether chimerical, arguing with perfect truth that there was no reason to believe that the law in question was anything more than an accidental coincidence,* and that it was thus utter madness to

---

* It has since been found to be broken in the case of the planet Neptune.

attempt reasoning upon it at all. The madmen, however, pursued their quest; and, after a long and interesting search, the first of the asteroids was discovered; and shortly afterwards, to the astonishment of everybody, a second, revolving in an orbit nearly coincident with that of the first. This remarkable departure from the established analogy of the whole solar system attracted universal attention; and when a third and a fourth asteroid had been discovered about the same place, Dr. Olbertz propounded the idea that the large planet which ought to have been found in this position had been, by some internal convulsion or by the shock of a comet, split into fragments—each of which was now pursuing its separate course as an independent orb about the great common centre of the system. This theory was at first almost universally received, being strikingly borne out by a remarkable fact with regard to the orbits of the then-discovered asteroids. If such a catastrophe occurred, the fragments would be hurled off in different directions and with different velocities, and would thus take up different orbits; but as the orbit of each would be ever the same, it follows that they would all at some period of their course pass again through the position from which they originally diverged. And this was found to be the case. There was a particular part of the heavens through which the four asteroids at one time or another passed, and which was therefore set down as having been the scene of the great original disruption. It was conjectured by some that the aërolites, or shooting stars, were small fragments from the same mass, which had been projected so far inwards towards the sun as to come within the range of the earth's attraction, and be deflected down to its surface. This latter hypothesis received a good deal of support, being at least as probable as that of Laplace, which refers the origin of these meteors to volcanoes in the moon, and holds that they are hurled forth from those lunar craters with force sufficient to reach the earth. But the explosion theory is now itself exploded. Many of the more recently discovered asteroids do not pass near the place of the supposed disruption; and therefore, as we have seen, can never have been at that spot at all. It is true that the perturbations caused by the other planets would by this time have partially affected their orbits; but the discrepancy seems too great to be accounted for in this way, and the theory has now been generally abandoned. The only other attempt to account for the phenomenon of the asteroids is based upon the great Nebular

Hypothesis of Laplace, which we shall explain hereafter.

These minor planets being all included within a belt of very moderate extent, it follows that large numbers of them will often be comparatively near together, and the appearance of the heavens at one of these will be peculiarly striking. Many bright planets will be scattered over every part of the firmament—some appearing as thin silver crescents like the new moon, some as half-moons, and others with fully illuminated discs; some so distant as to be indistinguishable from stars, and others surpassing the moon itself in magnitude and splendour; their orbits crossing and overlapping in every direction, and the planets thus circling in and out among each other as if in the mazes of some majestic dance,— some winging their flight far away to the most distant parts of their orbits beyond the sun, and others perhaps approaching so near as to fill half the firmament with their glorious blaze, and travelling along for days and weeks together, so near that their gigantic inhabitants might almost clear at a bold leap the airy gulf that separates their worlds from each other.

ASTROLOGY.

WE can scarcely turn away from the subject of the romance of planetary astronomy without alluding to the mysterious influence which those bodies of our system were for many ages supposed to exert on the affairs of men. The science of astrology—for a science, and a most elaborate science it was—comprehended, of course, the other heavenly bodies as well as the planets. But although the sun and moon are far more important luminaries than the planets, and although the stars incomparably exceed them in number, yet the simple regularity of their movements rendered them far less interesting to the astrologer than the "wanderers" of the nightly sky. To the ancients, unfurnished with the master-key of Copernicus, the motions of the planets, with their fitful loops and backward sweeps, appeared altogether arbitrary and irregular, and these orbs were therefore naturally selected as those most fitted to represent the varying turns of Fortune's wheel, and to preside over the changing lots of men, of nations, and of the human race.

The origin of astrology, or the fortelling of events from the configuration of the heavenly bodies, is lost in the mists of a remote antiquity, but it was undoubtedly practised by the old Egyptian magi, before the time of Moses. The father of the written science was the illustrious Ptolemy, whose astronomical researches seem to have been prosecuted mainly for astrological purposes, and whose elaborate work, the Tetrabiblos, is the text-book of all succeeding votaries of the science. According to him, the planet in the ascendant at the time of birth was the chief ruler of the character and fortunes of the "native," as the entrant on this world's stage was technically called. Mercury presided over the mental faculties, and literary and scientific occupations. He caused a desire of change—though in this respect his influence was less than that of the moon—and a love of travelling. Venus was a benefic planet, styled the Lesser Fortune. She tended to produce a mild and benevolent disposition, with an inclination to pleasure and amusement; and her favouring influence brought good fortune to the native in his or her relations with the other sex. Mars, on the other hand, was the Lesser

Infortune. His influence was not altogether evil, but he was decidedly risky, and needed to be well aspected by other planets to lead to any good. The man born under him was high-spirited, quarrelsome, and defiant of danger. The woman was probably a virago, or at the least what Ptolemy, if he had lived in a less favoured age, would have been familiar with as "strong-minded." Mars, of course, ruled over warlike pursuits, and also over such trades as were concerned with iron and steel.

Jupiter was regarded as far the most propitious of all the heavenly orbs, and styled the Greater Fortune. He ruled all high and dignified offices, especially the Church. The favoured mortal born under him might be expected to prove high-minded and honourable; charitable and devout; liberal, wise, just, and virtuous. Happy the kingdom ruled by a sovereign on whose birth he shone! English astrologers of the present day tell with pride that our gracious Queen was born when Jupiter rode high in the heavens, right upon the meridian. So, they say, was the Duke of Wellington; but as both the date and the place of his birth were uncertain, the astrologers must be as clever as Daniel—they can not only interpret the dream, but supply it when forgotten. The Greater Fortune smiled also, though less brightly, on the birth of the Prince of Wales.

Next him we have the grim and ill-omened Saturn, the Greater Infortune; "and justly," says Lilly, "does he merit the title, being the cause, under Providence, of much misery." Those born under him are gloomy and reserved in character; faithful, indeed, in friendships, but bitter and unforgiving towards an enemy. Failure, disease, disgrace, and danger beset the steps of the child of Saturn with frequent and terrible pit-falls. The only pieces of good luck that appear to be attributable to him are the gloomy ones of legacies; while his special favourites are sextons, undertakers, and mutes. Of Uranus, of course, Ptolemy tells us nothing, but modern astrologers think him on the whole malefic. He causes eccentricity and abruptness of manners; and whether he brings good or evil, it is always of some peculiar and unexpected kind. We cannot find how Neptune is regarded by the astrologers: probably they have not yet made up their minds about him. But we may hope for his credit that Adams and Le Verrier, to whom he owes so much, are watched over by him with special favour.

Although the ascendant planet is the chief element to be considered in Genethlialogy, as Ptolemy styles the science of nativities, its influence may be modified by its combination with other planets, or its position in the zodiac. Thus, while Mars in general begets military men, they must, if he be in the watery sign of Cancer or of Pisces, find vent for their fighting tastes in the navy. And so on, from the soldier and sailor, through the "tinker, tailor, ploughboy, and apothecary," down to the "thief," who is born under the moon, "afflicted by Mars." The tailor is the only one of the list we cannot trace. Probably, from his fractional character, he belongs to one of the asteroids.

The signs of the zodiac were supposed to have a good deal to do with personal appearance. Thus Pisces produced a short figure, pale and fleshy face, round shoulders, and a heavy gait; Taurus a well-set person, with broad face and thick neck; and so on. If parts of two signs occupied the ascendant together, a portion of the body would belong to one sign and the rest to another. Wild as the whole system of astrology is, it seems especially strange that the great philosophers of antiquity should have thought that human fortunes could be swayed, not merely by the constellations themselves, but by the arbitrary and fanciful names which men chose to assign to them.

Definite portions of human life were allotted to the different luminaries:—infancy to the Moon; childhood to Mercury; youth to Venus; the vigour of manhood to Mars; maturer age to Jupiter; and second childhood to the ominous Saturn. And lastly, the visible firmament was divided into twelve equal portions, meeting in the zenith. The first was the house of health; the second that of wealth; the third that of brothers and sisters, and also of short journies, the latter being probably put in to fill up the space if the former should be wanting; the fourth that of parents; the fifth that of children and of amusements; the sixth that of sickness; the seventh that of love and marriage; the eighth that of death; the ninth that of scientific pursuits and distant journies; the tenth that of trade or calling; the eleventh that of friends; the twelfth that of enemies. The connection of these houses with the rest of the system is, of course, obvious. Thus Saturn in the fifth house foretells misfortune with one's children; Mercury in the sixth house, mental disease; and Mars in the eighth house, a violent death.

Probably few persons have their horoscopes erected now-a-days, but we have before us that of the Prince of Wales, calculated at the time of his birth by Zadkiel, according to Ptolemy's rules. The Prince was born at forty-eight minutes past ten, on the morning of the 9th of November, 1841, at Buckingham Palace, lat. 51° 32′ N., long. 6′ W. The sign in the ascendant was Sagittarius, which, in Ptolemy's words, produces "a tall upright body, oval face, ruddy complexion (with a tendency to duskiness), chestnut hair, much beard, good eye, courteous, fair-conditioned, noble deportment, just, a lover of horses, accomplished, and deserving of respect." The Sun, being well aspected, prognosticated honours; and as he was in Cancer, in sextile with Mars, the Prince was to be partial to maritime affairs, and win naval glory. The house of wealth was occupied by Jupiter, aspected by Saturn; and this, as we have already seen, betokened "great wealth through inheritance"— a prognostication which, in spite of republican shoemakers and baronets, is not unlikely to come true. The house of marriage was unsettled by the conflicting influences of Venus, Mars, and Saturn, but fortunately the latter was to predominate, and the Prince, "after some trouble in his matrimonial speculations," was to marry a princess of high birth, and one not undeserving of his kindest and most affectionate attention. His marriage was to be expected in 1862. There are few other predictions of particular events; the one put forth with most confidence is that of an injury from a horse in May, 1870, when Saturn is exactly stationary in the ascending degree. Zadkiel says, however, that this evil might be guarded against by prudence, which we presume was done, as the accident did not come off. There was also danger of a blow on the left side of the head, near the ear; but it does not appear whether this was to be administered by the horse, or to be a separate accident. The house of sickness showed a predisposition to fever and to epileptic attacks. The position of Saturn in Capricorn betokened some loss or disaster to the native in one or other of the places specially ruled over by Capricorn; which we find from a table to be Brussels, India, Greece, Mexico, part of Persia, the Orkney Islands, and Oxford. We hope that the place indicated was the last of these, as if so the disaster is probably well over by this time, and was nothing more serious than some slight scrape with the authorities of Christ Church.

But while we have few particulars about the Prince's history, we are overwhelmed with information about his character. Each planet contributes an enormous list of characteristics, depending on its position and aspects at the moment of birth. When put together, they give the somewhat complex character which we subjoin. The infant Prince was to turn out "acute, affectionate, amiable, amorous, austere, avaricious, beneficent, benevolent, brave, brilliant, calculated for government, candid, careful of his person, careless, compassionate, courteous (twice over), delighting in eloquence, discreet, envious, fond of glory, fond of learning, fond of music, fond of poetry, fond of sport, fond of the arts and sciences, frank, full of expedients, generous (three times), gracious, honourable, hostile to crime, imperious, ingenious, inoffensive, joyous, just (twice), laborious, liberal, lofty, magnanimous, modest, noble, not easy to be understood, parsimonious, pious (twice), profound in opinion, prone to regret his acts, prudent, rash, religious, reverent, self-confident, sincere, singular in mode of thinking, strong, temperate, unreserved, unsteady, valuable in friendship, variable, versatile, violent, volatile, wily, and worthy." It will be seen that the good qualities largely predominate; the bad ones are due to Saturn, who of course must have his envious cut, but who is happily pretty well kept down by the cumulative influence of the propitious planets.

Zadkiel finishes thus:—"The square of Saturn to the Moon will add to the gloomy side of the picture, and give a tinge of melancholy at times to the native's character, and also a disposition to look at the dark side of things and lead him to despondency; nor will he be at all of a sanguine character, but cool and calculating, though occasionally rash. Yet, all things considered, though firm, and sometimes positive in opinion, this royal native, if he live to mount the throne, will sway the sceptre of these realms in moderation and justice, and be a pious and benevolent man, and a merciful sovereign."

God grant that it may be so, and that the life, so recently spared in answer to a nation's prayers, may, while crowned with every good and perfect gift itself, be blessed to the promotion of that nation's truest welfare.

THE MOON.

PASSING now from the planets to the other bodies of the solar system, we turn in the first place to our next door neighbour, the moon. While the interest with which we view the planets arises from their close analogy and consequently great probable similarity to ourselves, that attaching to the moon is caused mainly by its remarkable proximity to us, and the clear view which we accordingly have of its surface and configuration. It is, in fact, the only one among all the heavenly bodies of whose state and constitution we can ever hope to learn much by actual observation. With regard to the rest, we must for the most part reason from analogy alone, and hence we can seldom arrive at any results of which absolute certainty can be predicated. With the moon on the other hand, we have ocular demonstration; and though we do not know very much about it—not half so much as we should like—still what we *do* know we can be perfectly certain of, and that is a very great matter indeed.

Before we touch at all upon the vexed and difficult questions of the existence or non-existence of a lunar atmosphere, lunar seas, and lunar inhabitants, we may glance in the first place over those points of interest which depend simply upon the position and movements of the moon—points therefore in the determining of which there can be no difficulty, and about the results of which there can be no difference of opinion. The first ideas which the ancients conceived of the nature and constitution of the moon were very wide indeed of the truth. The old Chaldean astronomers supposed that it was a globe, one half of which was made of fire, and which, by revolving upon its axis, presented its different sides to us in succession. This idea accounted sufficiently well for the phases exhibited by it; it was, however, anything but a probable one in itself; and when Thales observed the fact that the bright portion of the moon is always that which is turned at once to the sun and to ourselves, the old hypothesis was at once exploded, and the true explanation—that our satellite shines with reflected solar light—came to be universally received. Next to the phases of the moon, the most noticeable point about the appearance which it presents to us, is the fact that the configuration of its surface is always the same.

From our earliest childhood that configuration, with its quaint resemblance to a human face, has been familiar to all of us; the large eyes and arched eyebrows of the "man in the moon," his irregular nose, and his long melancholy mouth are among our first recollections of the nightly sky. Nor is the idea only a tradition of the nursery. It is of the most venerable antiquity (though the ancients assigned to the moon's face a softer sex than we do); for we find in Plutarch the following quotation from a very early Greek poet, Agesianax, whose works are lost :—

πᾶσα μὲν ἥδε πέριξ πυρὶ λάμπεται, ἐν δ' ἄρα μέσσῃ
γλαυκότερον κυάνοιο φαείνεται ἠΰτε κούρος
ὄμμα καὶ ὑγρὰ μέτωπα τὸ δ' ἐρυθρὸν ἄντα ἔοικεν,—

lines which Amyot translates thus :—

> De feu luisant elle est environnée,
>   Tout à l'entour; la face enluminée
> D'une pucelle apparoit au milieu,
>   De qui l'œil semble être plus vert que bleu,—
> La joue un peu de rouge colorée.*

The earliest attempt at explaining the fact that the moon's surface presents a constant appearance to us, notwithstanding its revolution round us, is

---

> * Over the orb shines a resplendent light,
>   In midst of which a damsel's face is seen;
> Whose cheeks suffused display her blushes bright—
>   Her eye cerulean, or a pale sea-green.

found in Clearchus, a follower of Aristotle, who says: that "the moon must be the most beautiful and perfect mirror, in regard to smooth polish and lustre, in the world; for that in it we see to appear reflected the images and figures of our great continents and oceans." A little consideration sufficed to show that this hypothesis, besides its inherent improbability, was insufficient to account for the phenomenon in question; and astronomers were shut up to the conclusion that the moon rotates on its own axis in a period exactly equal to that of its revolution about the earth. This perfect agreement of two periods so independent of each other (in the case of the earth, for example, the angular motion about the axis is 365 times as rapid as that in the orbit) was long regarded as the most marvellous coincidence in the economy of nature; but a recent ingenious mechanical explanation, too difficult to be given here, has cleared away a good deal of its *à priori* improbability.*

---

* It has been shewn that if the two periods were originally at all nearly equal, the attraction of the earth on the protuberant parts of the moon would tend to bring them in time to exact equality. See Arago's *Astronomy*, vol. II. p. 283. Routh's *Rigid Dynamics*, p. 449. The same peculiarity has since been found to hold in the case of the satellites of Jupiter and one of those of Saturn.

The idea of Clearchus about the moon being a mirror was revived in a singular manner in the middle ages. Some pseudo-philosophers maintained the possibility of communicating between distant parts of the earth by reflection at the surface of its satellite. "Do we not," they said, "see objects sometimes reflected by mirrors, even in positions in which, by reason of the interposition of screens, we cannot see them direct? Accordingly, writing on paper, either in characters of the ordinary size, or magnified by optical arrangements, might be reflected up to the moon, and from thence be transmitted to some point of the earth. They might then be magnified by some means so as to become visible."

The necromancer Agrippa had the effrontery to maintain that he had actually communicated in this manner with the distant east. Nothing goes down so readily with the ignorant as a good round lie, coated with a flimsy varnish of science; and, accordingly, these marvellous asseverations were received with very general credence, and the scientific men of the day found considerable difficulty in combating them. The energy with which they controverted these fabrications made them perhaps the less ready to detect the grain of truth which lay concealed under the mass of fiction. The faint ashy light which irradiates the dark part of the lunar disc, and which produces the appearance familiarly known as "the old moon in the young moon's arms," was long a matter of discussion and debate among astronomers. Some supposed that the moon's surface was slightly self-luminous, others that its mass was partially transparent, and that the sun's rays penetrated to a small extent through it. But both these theories were disproved by the fact that in a total eclipse the ashy light was altogether wanting. It was reserved for an amateur—the painter Leonardo da Vinci—to suggest the real explanation; namely, that the illumination was produced by the sun's rays being reflected from the earth's surface to the moon's and back again to the earth. The astronomers gladly availed themselves of the suggestion, and being once put upon the right track, they had little difficulty in shewing that it presented a most perfect accordance with facts. It need not surprise us that the sun's light, even after two reflections, should remain bright enough to be discerned by the eye. When we consider the brilliant illumination which our own surface receives on a clear night from the full moon, it is evident that it must be quite possible for the lunar inhabitants to see their own light reflected back to them from us. And as the earth is so much larger than the moon, the effect will clearly be increased in proportion when we take the case of our light returned to us from them.

The general aspect of the heavens from the surface of the moon will not be very different from what it appears to us. The sun will be, of course, the great luminary in their firmament as it is in ours; and the great source of lunar—as of terrestrial—light and heat. The nightly revolutions of the stars also will be the same, and the place the moon itself occupies in our nocturnal firmament will be supplied to it by the earth, which will present the appearance of a splendid moon, thirteen times larger than the sun. Its revolution on its axis will present its different faces to the moon in rapid succession, and when our sky is free from clouds, the configuration of land and water on our surface will probably be clearly visible to the lunar inhabitants. They must know more of our circumpolar regions than we do, and could doubtless tell us whether there is open sea around the north pole; though, unless their telescopes are much more powerful than ours, they could not settle the question of the legendary Scotchman.

This great orb will appear immovably fixed in one particular part of the heavens, while the stars pass slowly beside and behind it. It will display the same phases, and cause and suffer the same eclipses, that our moon does. It is scarcely proper perhaps to speak of our suffering any eclipse from the moon at all; for the shadow of that body is so small, that it will never cover any large part of our surface, and will in fact appear only as a small black circle passing slowly across our disc. But solar eclipses on the other hand will be at the moon far more frequent and striking phenomena than they are here. From the large size of the body behind which the sun appears to pass, a total eclipse will sometimes last as long as four or five hours, during which time the whole surface of the moon will be plunged in midnight darkness.

In consequence of the slow rotation of the moon upon its axis, its day and night must each be a fortnight long; and as its year is just the same length as ours, each of its seasons must consist of only three days and three nights. But the distinction of the seasons will be much less there than on the earth, and will besides be almost entirely lost in the far greater difference between night and day. If the atmosphere at the other side of the moon be as attenuated as it is at that which is turned towards us, this fact, combined with the great length of time for which the sun continues above or below the horizon, will render the lunar days more scorching than the sirocco, and its nights colder than the frigid zone; and thus each of its long days will be in reality a summer, and every night a winter—the morning twilight spring, and the evening twilight autumn. The hemisphere turned towards the sun, or the part of the moon which appears bright to us, must have any moisture which it may contain dried up by his vertical beams; while on the other, or dark side, the ground must be frozen hard to the depth of several feet, the mountains covered with glaciers, and the seas blocked up with icebergs. At the very margin between the two hemispheres there will be a narrow temperate zone, which will of course move round the moon, as the latter turns round its axis and presents its different faces successively to the sun; and the only way in which we can see that life could be supported with comfort at the moon (supposing the atmospherical difficulty surmounted) would be by moving constantly round it, so as to keep always in this temperate zone. A queer Noah's Ark-like sight it would be to see the whole inhabitants of the moon, side by side, in a huge procession extending from pole to pole, and hurrying quickly round it at the rate of ten miles an hour—some riding, some driving, and some travelling in slow railway trains; beasts, wild and tame, galloping by their side, and all the birds of heaven flying along over their heads!

But this brings us to the great question whether the moon can really have any inhabitants or not. Of all the problems which the science of astronomy is called upon to answer, none perhaps is possessed of deeper or more general interest than that of the plurality or non-plurality of worlds. We have all often wondered, as we have gazed on the star-spangled sky, whether those distant orbs are teeming hives of busy life like our own, or whether all the inhabitants of the universe have been indeed collected upon this one tiny and insignificant ball. And as the moon is the only one of the heavenly bodies with regard to which there has ever been a chance of arriving at any positive and definite evidence upon this subject, it follows that upon it have been concentrated almost all the researches and arguments of astronomers on the point. We can fancy the eagerness with which Galileo first turned his tiny telescope to its mottled face, and his disappointment when he found himself unrewarded by any revelations of life at its surface. And as the instrument has received each fresh accession of magnifying power from his day to our own, every succeeding observer has felt the same anxiety and experienced the same disappointment. Kepler thought that he saw, in the regular circular valleys with which the moon's surface is so closely dotted, artificial excavations, under the sides of which the inhabitants sheltered themselves during their long and scorching days; but when he found on measurement how large the dimensions of some of these craters were, he was compelled to abandon the idea. Even within the last half-century, an eminent German observer, in using a new and powerful telescope, fancied that he had discovered a series of colossal fortifications in one part of the moon's surface, closely resembling the gigantic wall which the Chinese have erected against the outside barbarians. But these lunar ramparts could not stand against the tide of optical improvements, and the next big telescope showed them to be only basaltic formations, though of such singular regularity that their first observer might well be excused for attributing to them an artificial origin.

The fact is, that no satisfactory traces of inhabitants or of their works have ever been detected upon our satellite. The smallest space that can be distinctly seen with the best telescope at the surface of the moon is a circle of about a mile in diameter, and therefore no ordinary creation of human hands could be seen with sufficient clearness to place its character beyond doubt.

An old philosopher suggested, half in earnest and half in jest, a method of settling the point, which certainly possessed at least the merit of ingenuity. He argued that any race of rational beings must have discovered the leading principles of geometry, and would doubtless be aware that the square on the hypothenuse of a right-angled triangle is equal to the sum of the squares on its sides. He therefore suggested that a huge figure of the forty-seventh proposition of Euclid should be built on some great plain on the earth's surface. If the moon were inhabited by rational beings, they would be sure to recognize it as an old friend, and would doubtless divine that their terrestrial brethren were wishing to open communication with them. They would accordingly reply by the construction of some other important mathematical diagram—possibly, if their geometry is in advance of ours, they might send us down a method of squaring the circle. Thus we should at once have settled the existence of lunar inhabitants, and started a method of communicating with them. Probably our next move would have been to construct the figure of a man, to shew our new friends what we were like, and to hint that we should be glad to know something of them.

An improvement upon this suggestion was that enormous bonfires should be simultaneously kindled at points on the earth's surface forming the angles of a regular polygon. The symmetry of this phenomenon would strike the people in the moon with the idea of design, and suggest to them the existence of terrestrial inhabitants; and they would doubtless make known to us in return their own existence by some similar device. But neither of these experiments was tried, and neither of them is likely to be tried now; for in more recent periods some delicate investigations have thrown serious objections in the way of the inhabitant theory, by proving almost beyond a doubt the lack of water, and of all but an extremely attenuated atmosphere, on, at any rate, that side of the moon which is turned towards us. These conclusions were resisted as long as possible by Sir David Brewster, and other enthusiastic patrons of the "Selenites," or "men in the moon;" but latterly the weight of proof was becoming fairly too strong for them, and they were being obliged to take refuge in the somewhat unsatisfactory argument that after all the

Selenites might be so constituted that they could get on without either water or air.

But about fifteen years ago a great deal of light was thrown upon the subject from a new and most unexpected quarter—light that filled Sir David Brewster and his friends with extreme exultation, and carried confusion into the ranks of those who had too hastily triumphed over them. It was discovered, on the occasion of a certain eclipse, that the moon was no less than three seconds behind its time in touching the sun's disc!

What connection a fact like this has with the question of the moon's being inhabited or not it is not at first easy to see, but we shall find that it is really connected with it in the closest manner, and that in fact it clears out of the way all the objections which have ever been started against the moon's capability of supporting animal life at its surface. Of course it was not to be thought of for a moment that so great a discrepancy between fact and theory as that mentioned above should be allowed to remain unexplained. An express train on a journey of an hour's length would be granted at least a couple of minutes' grace, but not even three seconds could be allowed the moon after its long circuit of nearly a million miles. All the astronomers, both of this country and of the continent, were soon engaged on the question; and as it was proved that the observed irregularity was not due to the disturbing influence of any other body, they concluded ere long that it must arise from something anomalous in the figure or constitution of the moon itself. After an elaborate analysis, Professor Hansen, of Gotha, found that it could be accounted for only by supposing that the side of the moon nearest us was lighter than the other, and hence that its centre of gravity—or the point to which any object on its surface would be attracted—was not at its centre of figure, but considerably nearer the side of it which is always turned away from us. He calculates the distance between these centres to be nearly thirty-five miles, evidently a most important eccentricity, when we remember that the radius of the moon is little over a thousand miles. It must have been produced by some great internal convulsion after the moon assumed its solid state; but the forces required to produce this disruption are less than might at first sight appear necessary, owing to the fact that the force of gravitation, and the weight of matter, are six times less at the moon than with us.

From this peculiarity of the moon it follows that

any fluid substance at its surface, in its attempt to get as near as possible to the centre of gravity, must have flowed round to the other side, and taken up a position of equilibrium there—just as a drop of water let fall upon a smooth globe of any kind would trickle round it and hang suspended from the lowest point. In fact, we can readily see that the circumstances at the part of the moon nearest to the earth must be the same as at the summit of a mountain on our own globe more than two hundred miles high; and we know that at such an elevation as that, the atmosphere would be so rare as to be utterly indistinguishable. At the edges of the moon's visible disc, the conditions would be the same as at a considerable altitude on the sides of such a mountain. And as we find from actual observation that there really is a certain attenuated atmosphere at those parts of the moon's surface, we are led to the unquestionable conclusion that on its other side, which would correspond to the level surface of the earth, the atmosphere must attain a very considerable density, such as we have every reason to suppose would render it perfectly well fitted for the support of animal life.

The water difficulty is got over in a similar manner. It is true that when the seas flowed away from the side of the moon next us, large bodies of water would be left behind in lakes and in the depths of the ocean. But the withdrawal of the atmosphere would lead to their immediate evaporation, and as soon as they were converted into vapour, they too would be free to gravitate round to the other side. Indeed the visible disc of the moon presents every appearance of having been, in former ages, to a great extent under water. Its enormous level plains surrounded by lofty mountains, its huge basins or craters opening everywhere among the rocks, and the vast ravines dividing its mountain-chains, are altogether unlike our terrestrial scenery; but they are exactly similar to what we should see if all our oceans, lakes, and rivers were dried up, and their beds laid bare. In fact, we can scarcely doubt that there were formerly great bodies of water on this side of the moon; and if there were, it is equally certain that they must now be upon the other side.

We see, then, that we might have predicted *à priori* the absence of air and water from this side of the moon: we see also that there must be air, that there probably is water, and that there is no reason why there should not be inhabitants upon the other side. Astronomy thus clears away the difficulties it had itself raised in the way of the habitability of the moon; further than this it cannot go; no telescopes can ever pierce through the solid body of the moon and reveal the secrets of its further surface. Science has done all that can be expected at its hands when it has proved the complete possibility of a plurality of worlds, and the remainder of the question must be handed over to the inductive philosopher and to the natural theologian, to be judged of, as we have already said, by analogy, and by a consideration of what we otherwise know of the general economy of Providence. This part of the argument falls without our sphere—it belongs to the Religion of astronomy, and not to its Romance. The arguments, however, in favour of the plurality of worlds are patent to everyone, and each of us may arrive for himself at what conclusion he pleases upon it. For our part, we cannot bring ourselves to think that our own globe is the only inhabited one in the universe. Besides the *primâ facie* improbability that a small and insignificant planet should be in reality the most important body in creation, we think it is utterly impossible, on any common-sense grounds whatever, to believe that the larger and more distant orbs that spangle our firmament should have been created for our sakes at all. If their object were to afford us light, this purpose might have been far more effectually served by giving us another moon not a thousandth part as large as any of them; if it were to beautify our celestial scenery, this end, too, would have been equally attained by fixing some small luminous bodies within the limits even of our own atmosphere, instead of by placing these gigantic spheres at such incalculable distances from us. Nor is it probable that the Divine Architect should have created them for his own contemplation and that of the angelic hosts alone. It is contrary to the whole analogy of nature, and repugnant to all the ideas of the Divine wisdom and goodness which we have been accustomed to entertain, to think that these mighty orbs should have been framed for no other end than this. In all the economy of Nature we find nothing like waste of material or aimless expenditure of creative power; and while we see every blade of grass around us furnished with inhabitants and every drop of water teeming with a world of its own, it seems impossible to believe that those glorious stars should be in reality nothing more than so many waste and gloomy deserts.

Before we pass on to the next part of our subject, let us glance for a moment at a singular train of speculation suggested by a fact mentioned above—that, namely, of the eccentric gravitation at the moon. We have seen that any object at the surface of that body will be attracted not towards the centre, but to a point at a considerable distance from it. This eccentricity, though enough to bring all the lunar atmosphere and water round to the heavier side, will yet probably be insufficient to cause any serious practical inconvenience to the dwellers upon its surface; but if it were carried to a somewhat greater extent, the results would be very singular. The reason why the natural position of any object at the surface of the earth is an upright and not a slanting one, is that the centre of gravity being exactly beneath our feet, the direction of its attracting influence is of course perpendicular to the surface. But if the centre of gravity were removed somewhat to one side, its attraction would now be oblique, and all formerly upright objects, such as men, buildings, and trees, would be compelled to take up a slanting position in order to the preservation of their equilibrium, while any round and easily moveable body would immediately bound off in the direction of the new centre. Practically speaking, this state of affairs would be much the same as if the level surface were suddenly, and without any change on itself, transformed into a steep hill-side; for on such a surface, though at right-angles to the sea-level, we yet occupy a sloping position relatively to the face of the hill. But the eye would doubtless inform us that we were upon a level surface; and, in fact, in order to arrive at an idea of the matter, we must combine the appearance of a level plain with all the properties of a steep incline. At the part of the surface nearest the centre of gravity, this "slantindicular" state of things, as the Yankees would call it, would be especially singular. Though the surface all round would be evidently level, yet to whatever side we started off the feeling would be the same as if we began to ascend a steep hill; and while at the central spot a man would stand upright, when he walked away in any direction his head would seem to go faster than his feet, till he took up his natural inclined position. In fact, to use a slang expression in a strictly literal sense, it would be a regular case of "sloping off."

THE SUN.

THE next of the heavenly bodies which claims our attention is the great centre of the system itself. When seen with the naked eye, the uniform and dazzling brilliancy of the sun's disc prevents us from getting any idea of the configuration of his surface, as we can do in the case of the moon; and even when viewed through a telescope, the overpowering brightness of the greater part of it renders it impossible to distinguish anything of the surface from which that intense illumination proceeds: just as when the eye catches the glare from a fragment of glass lying in the sunshine, it sees only the light proceeding from it, while of the object itself it sees nothing. But fortunately for astronomers, the brightness of the sun's disc is not altogether uniform, and by a contemplation of those remarkable phenomena known as the solar spots, they have been able to arrive at an idea, and no doubt an approximately correct one, of the nature and constitution of this extraordinary luminary. The spots on the sun, though varying much in size and shape, yet in their general appearance partake very much of the same character. They consist of a black central spot or nucleus, surrounded by a well-defined fringe less dark in colour, which is known as the penumbra. Sometimes the nucleus is absent, more rarely the penumbra; but the great majority of solar spots show both. Round their edges there are generally seen small patches of light of intense brilliancy, surpassing even the ordinary radiance of the sun's disc: these are called faculæ. The spots are by no means permanent, but undergo changes, and often very rapid ones, in magnitude and position; while after a comparatively limited period they close up and disappear altogether, to be succeeded by new ones, and those again by others, in varying and never-ending succession.

The discovery of the spots on the sun was as stoutly resisted by the metaphysicians as those of Jupiter's satellites and the earth's orbital revolution. They could not in this case, like the superiors of the Inquisition, twist Scripture into contradiction with facts; but they took their ground on what with them was a still higher authority, the dictum of Aristotle. The illustrious Stagyrite had proclaimed the heavens incorruptible and immutable; it was beneath the dignity of the great orb of day to be affected by any physical changes such as our paltry planet is subject to. The Jesuit Scheiner, one of the earliest observers of the solar spots, was prevented by his provincial superior from publishing his results. "I have," says he, "read Aristotle's writings from end to end many times, and I can assure you that I have nowhere found in them anything similar to what you mention. Go, my son, and tranquillize yourself; be assured that what you take for spots in the sun are the faults of your glasses or your eyes." But other glasses and other eyes gave the same results, and only the most bigoted of Peripaticians could deny that their master's infallibility had received its death-blow.

The next discovery was a most important one. It was found that any individual spot, if it did not break up, moved across the sun's disc in a period of about fifteen days; that it then disappeared, and after an equal interval presented itself again on the opposite edge. This evidently pointed to one of two conclusions. Either the spots were solid bodies revolving round the sun, or else that luminary had himself a motion about an axis and carried his spots round with him. Scheiner and some other astronomers leant to the former hypothesis, but the latter was soon received as the correct one, and numerous theories were started as to what the nature of the spots really was. Galileo supposed them to be clouds. La Hire imagined that they were huge cinders from the burning body of the sun, rising to the surface of the fiery ocean which surrounded it, floating for a time upon it, and then being again engulfed within it, to rise a second time in another place. Derham and Wollaston referred them to volcanic agency, supposing that they were great clouds of smoke and scoriæ ejected from craters in a state of eruption, and that the faculæ consisted of flames and streams of molten lava. The least improbable hypothesis was that of Lalande, who believed them to be high mountains rising above the general solid surface of the sun, and sometimes covered, sometimes laid bare, as the tides and waves of the solar sea surged backwards and forwards around them.

But about a hundred years ago Dr. Wilson, of Glasgow, established, from simple optical considerations, the fact that the spots are depressions below the general luminous surface, and not eminences above it. It also came to be recognized that the rapidity of their fluctuations, and the gigantic scale on which they take place, are incompatible with anything but a gaseous state of existence. Accordingly the old idea of a luminous liquid ocean was discarded; and the following theory, propounded by Sir William Herschel, is now, in its main points, almost universally received. He supposes that the sun has two separate atmospherical strata, or rather gaseous envelopes of cloud-like consistency, both several thousand miles in thickness; the outer one— the photosphere, or source of the solar light and heat—being of some extraordinarily phosphorescent character, while the inner one is non-luminous in itself, but possessed of a highly reflective surface. Upon this theory the spots are caused by atmospherical agitations on a most enormous scale. A huge chasm, sometimes not less than fifty thousand miles in diameter, opens in the outer stratum, while a corresponding rift of lesser size in the inner one reveals the dark body of the sun itself behind. This accordingly constitutes the black nucleus of the spot, while the penumbra is caused by the light from the luminous atmosphere being reflected back to the eye from the surface of the inner one. If, as not infrequently happens, the rift in the photosphere does not extend through the inner envelope, the spot will be all penumbra without a nucleus. And when we have the rarer occurrence of the inner opening exceeding the outer in size, the whole spot will be uniformly black. The phenomenon of the faculæ, which Dr. Wilson showed to be great prominences or waves on the photosphere, is due to the piling up of the luminous matter thrown out from these gigantic chasms.

Not to enter further into details, which would be familiar to scientific readers and irksome to others, we may just say that this theory accounts in the most satisfactory manner not only for all the changing phenomena of the spots, but for other remarkable peculiarities which have been detected upon the sun's disc. Sir John Herschel has completed his illustrious father's theory by suggesting the probable physical cause of these convulsions in the solar atmosphere. He supposes them to be analogous to those terrible whirlwinds or rotatory storms which form so appalling a feature in the meteorology of our own tropical regions. Into this theory also it is impossible to enter in detail; but the whirlpool-like appearance of the spots, the situation of them all within a small distance from the sun's equator, their apparent rotation about an axis of their own, and the direction in which they move along the solar disc, all bear out Sir John Herschel's explanation in the fullest and most satisfactory manner. A *primâ facie* objection to his theory is that the terrestrial whirlwinds are caused by the great differences of temperature between the equatorial and other regions of the earth's surface, and that as the sun is the source of his own heat, there is no reason why such differences should exist in his case. But Herschel meets this objection by an ingenious answer. The sun has an ordinary non-luminous atmosphere of great extent, exterior to the photosphere; and his rotation on his axis will cause this atmosphere to bulge out considerably round the equator. And this greater depth of atmosphere, by retarding radiation from the equatorial regions, will give rise to the differences of temperature which the theory assumes.

Our knowledge of the physical constitution of the sun has been greatly increased within the last few years by the wonderful revelations of that most powerful engine of physical research, the spectroscope. A careful analysis of the solar spectrum formed by a prism, and a comparison of it with the spectra of terrestrial elements in a state of incandescence, reveal to us the presence in the solar atmosphere of many familiar substances, such as hydrogen, and the vapours of iron, sodium, and other metals. Line for line the solar spectrum agrees with the known peculiarities of elements which form constituents of our own globe, and we have the interesting fact established that the gorgeous parent of our system is, so to speak, bone of our bone and flesh of our flesh. The same powerful analysis, when extended to the stars, discloses similar results; and we are led to the inference that our own tiny globe, though such an insignificant fraction of the universe, contains, represented within its narrow bounds, all the materials of which that gorgeous system is built up. Unfortunately the spectroscope can tell us nothing of our own satellite, though it is so much the nearest and most distinctly visible of all the orbs of heaven. Moon-light is simply reflected sun-light; and hence its spectrum is, as we should expect, but a faint reproduction of the more brilliant solar one.

The spectroscope has lately been applied successfully to those singularly beautiful phenomena which accompany a total solar eclipse, and which are generally known as the rose-coloured protuberances. As soon as the sun's light is wholly cut off by the moon, cloud-like prominences, of a bright roseate hue, are seen projecting from its surface beyond the moon's edge; and occasionally traces of a layer of the same material are seen at their bases, which lead us to suppose that the whole sun is encompassed by a ring of this matter. Whether it is a distinct solar envelope, or only a part of the photosphere, is at present uncertain; but pending the settling of the doubt, it has received the specific name of the chromosphere. The spectroscope shows it to consist of incandescent gas, of which hydrogen is the chief constituent; and the rose-coloured protuberances are huge masses of this flaming substance, which have been hurled up into the solar atmosphere to a height, sometimes, of fifty or a hundred thousand miles above their ordinary bed.

Another interesting phenomenon which appears at the time of a total eclipse is the solar corona,— a great halo of light surrounding the darkened sun and stretching far out into space. This halo was at first supposed, naturally enough, to be the solar atmosphere, lighted up by the sun's rays streaming through it and imparting to it a portion of his own effulgence. But here again the spectroscope comes to our aid. It tells us the degree of pressure to which the incandescent hydrogen composing the rose-coloured protuberances is subjected, and shows the impossibility of their being burdened by such an enormous atmosphere as the whole corona would represent. The progress of modern science has left little doubt as to its real nature. We have learnt that the whole solar system is traversed by numberless tiny planetoids, some moving singly, others in small clusters, and others in enormous groups, containing countless myriads of these little units. These aërolites pursue their proper paths about the sun as truly as the largest bodies of the system, save when they get entangled in the atmosphere of our own or any of the other planets. When this is the case, the sudden checking of their enormous velocities by the resistance of the air reduces them instantly to a state of incandescence, and we see them flashing across our firmament as shooting stars, the next moment to be dissipated into vapour. The periodical meteoric showers of August and November are caused by our orbit carrying us, at those periods of the year, right through great clusters of these aërolites. It has been estimated that not less than a hundred thousand million of them are annually caught by our atmosphere; and when we consider the comparative smallness of the ring which we traverse, we can see that the absolute number of the meteorolites belonging to our system must be something incomparably exceeding the highest flights of human calculation. In the immediate neighbourhood of the sun, where his attraction exercises the most direct and potent influence, they will be found in special abundance;

and it is to the fact of their existence that we must look for an explanation of the corona, and perhaps of yet greater and more interesting mysteries of our system. The corona is simply the sun-light reflected from their surfaces, as it is from the discs of the moon and planets. For a vast distance round the sun the whole firmament is powdered with them as thick as hailstones, and the reflection from them produces a continuous luminous glow, lost indeed in the overpowering brightness of ordinary sun-light, but shining out with exquisite lustre when his direct beams are cut off from us.*

These meteorolites have played a most prominent part in the scientific speculations of the last twenty or thirty years. The meteoric theory of the sun's heat was first propounded by Dr. Mayer, a German physician of great scientific attainments, and was warmly espoused and worked out by Sir William Thomson. There can be no doubt that the sun is constantly receiving great accessions of heat from the meteoric fragments which compose his corona. Countless myriads of them rain into his atmosphere every instant, with a force sufficient to convert their solid

---

* It appears probable that it is also, under favourable circumstances, seen after sunset in the form of the zodiacal light.

mass into a puff of vapour; and it was for some time thought that the heat derived from these terrific impacts might keep the solar envelope ever ablaze with undiminished intensity. No grander or more striking theory has ever been propounded in the history of astronomy. That the mighty centre of our system should recruit his marvellous expenditure of energy from the tiniest of his satellites; that these fragments, each so insignificant in itself, should collectively supply light and heat and life to the great Sun himself, and through him to all his attendant orbs; that it should be through their agency that the Creator of the universe has ordained that all His creatures should live and move and have their being,—is one of the most striking conceptions that can possibly be imagined. But more recent observations have led Sir William Thomson to a modification of his theory. He has calculated that if the meteoric shower were sufficiently heavy to make up for the sun's whole expenditure of heat, the matter of the corona must be so dense as seriously to perturb the orbits of certain comets which pass very close to his surface,—a result which is found not to be the case. But the meteoric theory is only thrown back a step. If the sun's mass were originally formed, as is not at all improbable, by the agglomeration of these particles, Sir William Thomson has calculated that the heat generated by their thus falling together would be sufficient to account for a supply of twenty million years of solar heat at the present rate of emission. And thus though the meteors are not sufficient to maintain the energy of our system unimpaired, they may yet have been the original store-house from which all that energy was derived.

The fact, now placed beyond doubt, that the sun's heat is gradually wasting away, naturally leads us to cast a glance into the future. Far, very far, distant the time must be; long before it comes, in all probability, the firmament will have been rolled together as a scroll, and the old heavens and the old earth will have passed away. But if the economy of our system be spared long enough, the day must come when the sun with age has become wan ; when the matter of the corona has been all drawn in and used up without avail; when the lavish luxuriance with which he has showered abroad his light and heat has finally exhausted all his stores. He has still power, aided by the resisting medium,* to drag his satellites one by one down upon his surface;

---

* See the Chapter on Laplace's Nebular Hypothesis.

and the shock of each successive impact will, for a brief period, give him a fresh tenure of life. When the earth crashes into the sun,* it will supply him with a store of heat for nearly a century, while Jupiter's larger mass will extend the period by thirty thousand years. But when the last of the planets is swallowed up, the sun's energies will rapidly die out, and a deep and deathly gloom gather around nature's grave. Looking into the ages of a future eternity, we can see nothing but a cold and burnt-out mass remaining of that glorious orb, which went forth in the morning of time, joyful as a bridegroom from his chamber, and rejoicing as a strong man to run a race.†

---

* In the chapter above referred to we have depicted the earth as falling into the sun, still an intensely heated body. But it is scarcely possible to say whether such a catastrophe, should it ever occur, would not find the sun already a comparatively cold mass.

† The scope of these pages does not admit of our entering at greater length into the interesting and important subject of recent solar research. Upon this point our readers must of course refer to larger and more systematic works, or to the original papers scattered over various scientific periodicals. A clear and interesting discussion of solar heat will be found at the close of Professor Tyndall's work on Heat; while the invaluable spectroscopic revelations of Kirchoff, Lockyer, Huggins, and others, which have added so vastly to our knowledge, not only of the sun but of other astronomical bodies, will be found well described in Roscoe's Spectrum Analysis and Proctor's work on the Sun.

THE COMETS.

In a sort of "debateable territory" between our own solar system and the infinite stellar universe around, we come upon those erratic and anomalous bodies—the comets; some of which have accidentally become permanent attendants upon our sun; others have only paid it a single casual visit in the course of their wanderings through space, and are not likely again to come within the range of its attracting influence; while countless millions are doubtless scattered throughout the realms of the infinite, whose existence will never be revealed to human ken at all. The extraordinary appearance and anomalous character of these meteors, the apparent irregularity of their movements, the suddenness with which they blaze into the firmament, the gigantic trains of light which they throw out as they near the sun, the frightful velocity with which they whirl round that body, and the sudden diminution of their glory as they recede from it, till they seem to be extinguished in the primæval darkness from which they emerged, —all these circumstances, combined with the mystery in which their real nature is shrouded, have caused these knights-errant of astronomy to be regarded at all times with the deepest interest, generally not unmixed with superstitious dread. In fact, for ages they were hailed by the universal consent of all classes of the community in the light of portents. One was believed to have portended the birth of Mithridates, another the assassination of Julius Cæsar, a third the great plague of 1310. One of the most remarkable on record made its appearance at the time of the Saracenic invasion of Christendom. As the hosts of the Crescent swept on in their irresistible course, the comet waxed brighter and brighter, till at last, as the Caliph Mahmoud laid siege to Constantinople, it filled half the sky with its splendour, and hung night after night over the doomed city in the guise of a blazing scimetar. The Pope scarcely knew at first whether to pray to, or to curse it, but adopting the latter course as more congenial to a true Catholic spirit, he fulminated the thunders of the church against it, and in the same bull excommunicated both Moslems and comet.

Though the superstitious terrors which used to greet the appearance of a comet have now for the most part passed away, yet the mystery which involved them, and which in a great measure gave rise to those terrors, has by no means been altogether dispelled; for our ideas of the constituent matter of their several parts can scarcely be regarded as more than conjectural. Most of the comets which are visible to the naked eye consist of two separate and well-defined parts—a round nucleus or head, and a long filmy train of light, the 'coma' or 'hair,' from which comets take their name, and which is known in popular parlance as the tail, though this appellation is not altogether a correct one, for, unlike the sheep in the nursery rhyme, comets do not always "carry their tails behind them," but quite as often in front of them. In former times, indeed, the hairy appendage, when worn in front, was called a beard; but the distinction has now ceased to be drawn, and front-hair and back-hair alike go by the name of tail. The nucleus, or apparently denser portion of the comet, diminishes considerably in size when near the sun; while the tail, on the other hand, undergoes a most remarkable augmentation, its length sometimes increasing no less than a million of miles in a single hour.

If, as was universally assumed until within the last few years, the tail of a comet be composed of continuous matter, it must unquestionably consist of some extremely rare gaseous substance, incalculably lighter than our own atmosphere. This is inferred from their inappreciably small weight. Though a comet is often many million times larger than the sun, its mass is yet so insignificant that the most delicate tests fail to detect it at all. Many of us must remember also how in the last great comet, that of 1858, the star Arcturus shone with undiminished brilliancy through the thickest part of the tail. The lightest cloud wreath would have concealed it altogether, and yet the filmy texture of the comet, though millions of miles in depth, could hardly dim its lustre. The enormous increase in the size of a comet's tail when near the sun is simply explained, on the gaseous hypothesis, by attributing it to expansion by heat. The great rarity of its mass would render it extremely sensitive to variations of temperature, and account readily enough for its rapid expansion when near the sun, and contraction when receding from it. The nucleus would consist of a material similar to, though somewhat denser than, the tail; and its shrinking at perihelion would probably be due to the conversion of its outer parts into invisible vapour by the action of the solar rays.

But a novel and most ingenious theory, entirely revolutionizing all our former ideas about comets, has been very recently started, and has met with much acceptance among astronomers. When the showers of falling stars in August and November had been discovered to be caused by the passage of the earth through great clusters of meteoric stones, the paths of those clusters about the sun were investigated, and were discovered to be the same as the orbits of two known periodic comets. This extraordinary coincidence appeared too remarkable to be the work of chance, and the bold idea was conceived that these phenomena, apparently so different, might in reality be identical. And on examination, the great *à priori* improbabilities of the hypothesis are seen to disappear in the most wonderful manner, and it is found to afford not only a possible, but a highly probable, theory of the nature and constitution of the cometary bodies. Every comet is supposed to consist of a vast assemblage of distinct solid masses—stones, rocks, and lumps of metal—flying together through space, and rendered visible, in favourable positions, by the sunlight reflected from them. The smallness of their aggregate mass, and the fact of their not eclipsing any heavenly bodies which pass behind them, are as well accounted for by this supposed discontinuity of their material as by assuming them to be gaseous. So are the sudden changes in shape and size of the tail. It will not be visible except when the comet is in such a position as to turn a sort of flat edge towards us, so that we can look at once through a great depth of its mass. For the reflection from each elementary fragment will be so slight, that it will be only when an enormous number are ranged along the same line of vision that their aggregate light will be sufficient to affect the eye. To borrow a felicitous illustration from Professor Tait, we may see the same thing represented in miniature by the flight of a flock of sea-birds. Great numbers of them often fly about, approximately in one plane; and if they are at such a distance as not to be discernible singly, they will be equally invisible when their plane has its face turned towards us. But when a sudden sweep brings them into the plane of our vision, so that we get a number of them in one line, they start into sight at once, as a black streak against the face of the sky.

The nucleus is believed, from recent spectroscopic observations, to consist wholly or mainly of incandescent gas. But the explanation of this necessitates no addition to the meteorolite theory. In the head of a comet, where its component fragments are crowded most closely together, there must be very frequent and violent collisions between them; and the heat generated by these impacts will convert them into vapour, just as we know to be the case when they strike our own atmosphere. Unfortunately no large and bright comet, which could be observed in a thoroughly satisfactory manner, has appeared since the invention of the spectroscope. We may hope that the next one which visits our skies will settle the vexed question of the constitution of the tail. The polariscope tells us that it shines by reflected sunlight, but whether partially or wholly it is unable to decide. If there be any self-luminosity, the spectroscope will reveal it at once, and will show also whether it emanates from solid or gaseous matter. And even if the light is found to be all sunlight, there is reason to hope that if the cometary matter be gaseous it will modify the spectrum sufficiently to make us aware of that fact. On the other hand, if the tail consist only of solid particles, the spectrum will be purely and simply an enfeebled solar one, with all its lines absolutely unmodified. Meanwhile the scale of evidence seems to incline in favour of the newer theory. Professor Tait maintains that certain facts which have never been satisfactorily explained on the old hypothesis can be successfully grappled with by the new one.

The paths pursued by the comets are very various indeed. Many of them, like the planets, move in ellipses round the sun, some traversing their orbits in three or four years, while others roam so far away that many centuries elapse before they again revisit our neighbourhood. A great number, however, only circle once round the sun, and never return to it again. The orbits into which, in accordance with the law of gravitation, they are bent, are so inconceivably long, that before they can reach the farther parts of them they come within the attracting influence of other stars, and are drawn off to pursue new orbits around new centres; and in this way a comet may wander through the universe for countless ages, seeking rest and finding none, till at last some star seizes it with a firmer grasp than the rest, compels it into a smaller orbit, and thus secures it as a perpetual attendant upon itself.

The question has often been asked, whether there is any likelihood of one of these nomadic bodies coming into collision with our own earth, and if this event *did* take place, what its effect would be upon us. On the meteoric theory the answer to this question is a startling one. The earth *does* come into collision with a comet regularly twice every year, and the result is simply a shower of shooting stars, more or less numerous and brilliant, according to the density of the portion of the tail which we encounter. But we must by no means conclude that every such collision would be attended by equally harmless consequences. Fortunately the comets which we encounter are composed of small and widely-scattered fragments, but many will probably consist of far larger masses more densely crowded together. Numbers of meteoric stones are too large to be converted into vapour during their passage through our atmosphere, and reach the ground in a solid, but red-hot, state. Several of these are on record which weighed more than a hundred pounds, and one, which fell in Spain in 1810, measured thirty inches in length and weighed three-quarters of a ton. An encounter with a comet composed of such masses as this would be a frightful ordeal for the earth to pass through. Its whole surface would be bombarded for some minutes or hours with great lumps of red-hot rock, which would burn and destroy everything upon which they fell. The only chance of the human race surviving such a catastrophe would lie in the fact that astronomers would probably foresee its advent, and warn everyone to take refuge in cellars or under bomb-proof casements, from which they would emerge, after the storm was over, to find all around them a mass of blazing ruins.

Nor is even this the worst form which a collision with one of these dangerous bodies might take. An encounter with the head of a comet would be a far more destructive event than a passage through its tail. And such an event, though extremely improbable, is yet a perfectly possible occurrence. Arago estimated the chances against any particular cometary nucleus striking the earth to be about three hundred millions to one; but still these chances, great as they are, must be by no means confounded with certainty; and indeed we find that on more than one occasion such a collision has very nearly happened. The nucleus of the comet of 1832, for instance, would have struck the earth if it had only been a month sooner. The consequences of such a catastrophe are almost too horrible to be contemplated. If, as is now generally believed, the nucleus consists of incandescent gas, we should find ourselves plunged in an instant into a mass of blazing vapour, which would scorch every trace of life off the earth's surface, and, not impossibly, dissipate its solid mass in smoke. If, on the other hand, as several astronomers of note have believed, certain cometary nuclei are composed of one solid mass, the results of a collision would be little less disastrous. Laplace describes them thus:—"It is easy to represent the effects of the shock produced by the earth's encountering a comet. The axis and the motion of rotation changed; the waters abandoning their former position to precipitate themselves towards the new equator; a great part of men and animals whelmed in a universal deluge, or destroyed by the violent shock imparted to the terrestrial globe; entire species annihilated; all the monuments of human industry overthrown—such are the disasters which a shock of a comet would necessarily produce." And even if, returning to the old hypothesis, we suppose ourselves coming into contact with a purely gaseous comet, though no mechanical shock of the above nature would be experienced, the chemical consequences might yet be equally fatal. Whatever the cometary material may be, it is not likely that it will be the same as that which composes our own atmospheric air; and, as our lungs are not adapted for the inhalation of any other kind of gas, the probable effect of the intermingling of our atmosphere with the substance of a comet would be at once to render the former utterly unfitted for the support of animal life.

LAPLACE'S NEBULAR HYPOTHESIS.

WE have now completed our survey of the great system of which we ourselves form a part. Sun, planets, satellites, and comets—all the elements of the solar system—have successively passed before us; and the only heavenly orbs we have left to consider are the more distant stellar ones, which are so far removed from our own immediate ken. But before we proceed to visit those distant realms, we must glance for a moment at Laplace's great theory of the origin of our system, one of the grandest and most magnificent speculations which it has ever entered into the heart of man to conceive. It is true that Newton's discovery of the law of gravitation furnished us with a key to much that was dark and inexplicable before; it reduced the motions of the planets to harmonious symmetry, and replaced the elaborate eccentrics, cycles, and epicycles of the ancients by simpler and more familiar curves. That law, as all our readers are aware, explains why the planets move in ellipses, and accounts for their different periods and ever-varying velocities; but there are questions which it does *not* answer—such as why all those elliptical orbits are so nearly circular, and why the planets and satellites all revolve round their primaries and rotate on their axes in the same direction,—namely, from west to east. On these points Newton's law throws no light; the solar system would be equally possible, and, with certain limitations, equally stable, and equally fitted for the support of life if this remarkable uniformity did not exist. To what then are we to attribute it? Is it likely to have been a direct and arbitrary exercise of the Creative will, or a less direct result of that power, working by natural means from some prior form of existence? Were the sun and planets called suddenly into being in their present forms and with their present motions, or were they developed by slow and gradual steps from some simpler original creation? Science tells us that both hypotheses would accord sufficiently well with known present facts; reverential thought pronounces neither of them inconsistent with the loftiest views of the Divine power and wisdom. But the latter is certainly the more interesting and attractive to us, and, we may perhaps add, apparently the more consonant with what we see of the Almighty's working in the lesser world around us. We see the perfect man developed from a helpless babe; we see the loftiest tree developed from a tiny bud; and there is nothing incongruous in supposing that our glorious system itself may have sprung from some vast but equally simple germ. This is a subject which must be handled with humility and diffidence; for we are leaving the regions of mathematics and entering upon those of uncertain speculation; we are treading on sacred ground in seeking to enter into the counsels of the Great Architect of the Universe. We are endeavouring to go deeper than the laws of nature will take us; we are seeking for a key to the mysteries of our system in the probable circumstances of its creation. For this it is that the bold and lofty speculation of Laplace seeks to do; it reaches back into the unfathomable ages of a past eternity, and takes its stand beside the Almighty Author of all things at the first exercise of his creative fiat, when the foundations of the earth and the heavens were laid, when the morning-stars sang together, and all the sons of God shouted for joy.

The Nebular Hypothesis is briefly and poetically summed up as follows by the Poet Laureate in "The Princess:"—

> "This world was once a fluid haze of light,
> Till toward the centre set the starry tides,
> And eddied into suns, that wheeling cast
> The planets."

Laplace supposes that the first great act of creation—possibly the only one which strictly merits the name, the only one in which things that are seen were made not of things that do appear—was the calling into existence everywhere throughout the infinite regions of space a huge, chaotic, and nebulous mass of matter, such as was supposed to form the substance of the cometary bodies.* This is the only hypothetical step in the whole argument, all the rest following from it, as we shall see, by a course of the most rigid deduction. And against this hypothesis we think no reasonable objection can be urged. It accords with the language of Scripture—"the earth was without form and void, and darkness was upon the face of the deep." It harmonizes with the other sciences; the geologist, in particular,

---

* We give in the text Laplace's own theory unaltered. Those who accept the modern hypothesis of the nature of comets would have to replace his "fluid haze of light" by a vast cloud of meteoric stones. But this supposition would not invalidate the theory; it would accord equally well with the rest of Laplace's speculation; and every step in the after development would be the same in either case.

having often to suppose the existence at a former period of a state of things springing from some such origin as this. Nor is it only to the man of science that the idea commends itself; it is a favourite theme with the poets. We find it in the pages of Hesiod, of Ovid, and of Dante. Milton gives it expression thus:—

> "A hoary deep, a dark
> Illimitable ocean, without bound,
> Without dimension, where length, breadth, and height,
> And time, and place, are lost; where eldest Night
> And Chaos, ancestors of Nature, hold
> Eternal anarchy, amidst the noise
> Of endless wars, and by confusion stand.
>   A wild abyss,
> The womb of Nature, and perhaps her grave,
> Of neither sea, nor shore, nor air, nor fire,
> But all these in their pregnant causes mixed
> Confus'dly....... Chaos umpire sits,
> And by decision more embroils the fray
> By which he reigns: next him, high arbiter,
> Chance governs all."

But the Spirit of God moves upon the face of the waters, and the first elements of order begin to emerge from primeval chaos.

As all the particles of this nebulous mass would exert a mutual attracting influence upon each other, it follows, in accordance with the law of gravitation, that they would begin to settle down and condense gradually around certain centres, the matter at which, from the intestinal workings of the whole, had become denser than the general mass around. And each of the nuclei thus formed contains the embryo of a separate sun; in each void chaotic mass the eye of the philosopher can already detect the germ not only of the great central orb, but of all its gorgeous band of attendants—planets and asteroids, satellites and rings. Let us follow the history of one of these nebulæ—say the parent of our own system— and trace the steps of its gradual elimination from chaos and conversion into the glorious cosmos which we now behold it.

The particles of this mass will all, of course, gravitate towards the centre, and a spherical form will thus be assumed by the whole body; but as the particles will approach the centre from opposite sides and with different velocities, a motion of rotation round an axis will necessarily be generated—slow at first, but rapidly increasing in velocity. As the ball condenses more and more, and shrinks into smaller bulk, it will, by a familiar mechanical law, spin round faster and faster; till at length the centrifugal force at the equatorial parts will overbalance the attraction of gravity, and a ring of surface-matter will detach itself from the general mass, and remain poised in mid-air behind as the ball within shrinks further and further away from it. The same process will be repeated over and over again, until at last the central mass becomes sufficiently solidified to resist any further separation of its parts. Now if we look at the case of any of these rings, we shall find that the form it eventually assumes will be different in different cases. If its material, as it is detached from the central mass, should happen to be of extremely uniform consistency throughout, and to be poised with extreme accuracy about its centre, it might possibly cool down and solidify in the ring-shape;* but the chances against this are so great that we should expect it to be a very rare phenomenon indeed. If the density of the ring were at all irregular, it would inevitably split up into fragments, as the cohesion of its parts would be very slight. The largest of these fragments would, by its superior attraction, assume the others

---

* This part of Laplace's theory must be somewhat modified to render it consonant with the discoveries of modern science. It has been shown that a ring, such as Saturn's, could not possibly exist in a solid state, but that it must be composed of separate fragments, or meteorolites, such as constitute the matter of the solar corona, and probably of the comets. But the theory readily adapts itself to the explanation of such a phenomenon. The ring would be detached from its primary in a viscous state; and as it would be impossible for it to solidify as a whole, it would break up into small fragments which would solidify separately and move in nearly coincident orbits, thus preserving the general form of a ring, although not one composed of continuous matter.

into its mass, and the whole would solidify into one globe of considerable size, except in the rare case of all the fragments happening to be about the same magnitude, when they would continue separate, and revolving round their primary in very nearly coincident orbits. Of course the planets, as they were thrown off from the sun, would proceed, in turn, to develope satellites of their own in the same manner and with the same possible varieties.

Now all of these results we find actually occurring in nature. The perfect rings we have in the case of Saturn; the groups of small planets near together are represented by the asteroids; while in every other case we meet with the arrangement which we have seen would be the most likely to occur—a large satellite revolving round its primary, and situated at a considerable distance from any of the others. All the peculiarities of planetary motion, too, are accounted for by this theory. The planes in which the planets move are nearly coincident with the plane of the sun's equator, because the matter of which they are composed is thrown off from the tropical parts of that body. Their orbits are nearly circular, because such would be the motion of their particles while yet in the ring condition. And the direction in which they revolve is the same as that in which the sun turns on its axis, because they would acquire an impulse in that direction before they parted company with it.*

The nature and origin of the comets are also easily explained on this hypothesis. When the original nebulous material of the universe began to gravitate towards its several centres, large masses of it seem to have been in many cases left behind, too evenly balanced between the different attracting influences to yield to any one of them. But as their position would be one of unstable equilibrium, in the course of time the attraction of some one or other of the centres would come to predominate, and the filmy and uncondensed mass would gradually yield to its sway, and descend towards the controlling orb. Another very mysterious point which the nebular hypothesis accounts for, is the existence of a resisting medium around the sun, and extending to a considerable distance from it. The existence of such a medium is now undoubted. Encke's comet, which

---

* The theory also accounts for the fact of the planets rotating about their axes in the same direction as that of their orbital revolution. When a fragment was detached from the ring, its outer particles would have a greater velocity in the general direction of motion than its inner ones, and there would be on the whole a moment of momentum in that direction about the centre of gravity.

possesses the smallest orbit of any connected with our system, is sensibly drawing nearer and nearer to the sun at every revolution; and this fact cannot be accounted for in any other way than by supposing this medium to exist. On Laplace's hypothesis its origin is readily explained; it must evidently be part of the original solar nebula, which, from its extreme rarity, has never undergone condensation at all.* The terrible part which this resisting medium may be destined to play in the great drama of the universe is indicated but too plainly by the effect it has already begun to produce on the comet which comes most immediately under its influence. Slackening by sure though imperceptible degrees the speed of every planet and comet in the system, and thus stealing away their power to resist the sun's attracting force, it will, by its insensible influence, bring them all in time within that orb's resistless grasp, till one by one they drop through his fiery atmosphere and sink to rest upon his surface. Thus, to quote the words of Professor Nichol, one of the warmest advocates of the nebular hypothesis—"The first indefinite germs of the great organization of the universe, provision

---

* On the more recent theory, it will probably be simply the matter composing the solar corona.

for its long existence, and finally its shroud, are all involved in that master conception from which Laplace endeavours to survey the mechanisms amid which we are. Not in confusion, however, shall this majestic system finally pass away—not with the jar and confused voice of ruin, but even in its own quiet and majestical order, like the flower which, having adorned a speck of earth, lets drop its leaves when its work is done and falls back obediently on its mother's bosom."

The terms in which the final destruction of our earth is spoken of in Scripture, and the comparatively short existence which seems to be in Providence destined for it, render it pretty certain that this globe at least will not meet with its doom in the above-mentioned manner. But as such an event is not only a perfectly possible one in the economy of nature, but an absolute certainty supposing that the resisting medium were allowed time enough to do its work,[*] it may not be out of place to pause for a moment and consider what is involved in such a catastrophe. Let us think, for example, what the

---

[*] The fact of the gradual dissipation of the energy of our system, established by Sir William Thomson, points also to the final destruction of the earth, and would tend to hasten the catastrophe we describe.

case would be with our own earth, if no speedier destruction were to come upon it from some yet unanticipated and possibly miraculous cause. Many centuries no doubt—it may be many milleniums—would elapse before the most delicate observations could reveal the working of the mysterious agent. But at length some astronomer detects a minute change in the elements of the earth's orbit which cannot be accounted for by any of the ordinary perturbations, and he is compelled to the belief that the resisting medium is beginning perceptibly to influence the planet. This discovery, when publicly announced, could not fail seriously to impress the most thoughtless of hearers. The first step has been taken by the earth on its way to a doom as fearful as the imagination can paint and as inevitable as the unchanging laws of nature can make it. Still generation after generation passes away; the end is—visibly—no nearer, and but for the figures of the astronomers the whole thing might be denounced as an idle fable. But not the less surely does the unseen destroyer fulfil his mission; and in time the effects of his work become palpable to every eye. The sun's disc is perceptibly enlarged, the intensity of his light and heat are increased, the length of the year is diminished. At first the change of climate is a pleasant and grateful one, except between the tropics, and even there it is not so marked as to be very severely felt. But slowly and surely the influence becomes more potent, and when we look again some ages later, the face of the intertropical regions is scarcely recognizable. The rich vale of the Nile, the fertile plains of the Ganges, the cotton plantations of the south, have disappeared; the sandy deserts of Africa and Asia have extended their bounds and stretch without an oasis far on either side of the equator. The inhabitants retreat, some to the north and some to the south, but the fiery belt between steadily pursues them, and mile after mile, league after league, falls under its devastating sway. Some ages more pass away, and when we look again the vineyards of Spain, the olive-groves of Italy, the fig-gardens of Turkey, are gone; their cities yet stand with all their splendid palaces, their gorgeous temples, but they are like Tadmor in the wilderness—cities without inhabitants. Look again, and Mont Blanc has lost his diadem of snow and rears his head, a bare cone of granite, above the dry and rocky table-land which was once the Mer de Glace. Look again, and our own land has, in its turn, become a burning desert. And now the whole inhabitants of the globe are collected in two narrow circles around either pole. The ice and the snow have disappeared, and the frozen plains of Greenland and Labrador teem with tropical vegetation. But the narrowed limits of the habitable earth can no longer support this vast increase of population, and famine begins to mow down its victims by millions. Now, indeed, the end of all life on the earth draws on apace. The resisting medium, from the increased proximity to the sun, grows rapidly much denser, and its effect is proportionately increased. The heat and drought become more and more insupportable. Rain and dew fall no longer. All springs of water fail, and the rivers dwindle down to streamlets, and trickle slowly over their stony beds. And now scarcity of water is added to scarcity of food. Those who escape from the famine perish by the drought, and those who escape from the drought are reserved for a fate more awful yet. For a time, indeed, the few remaining inhabitants of the earth are partially screened from the overwhelming power of the sun by a dense canopy of clouds. From the excessive evaporation, thick columns of mist are constantly rising from the surface of every lake and every sea, and forming into dense banks of cloud, which hang like a funeral pall over the dying earth. But soon the sun scorches up these vapour-banks and dissipates them into space as fast as they can be formed by evaporation. Then the fiery orb shines out in unutterable splendour without the lightest cloud-wreath to interpose between himself and his victims. Then, truly, the heavens become as iron and the earth as brass. Then the last denizens of the world are stricken down and consumed, the last traces of organic life are blotted from its surface. How different the "last man" here from Campbell's picture of

> "The last of human race,
> Who shall creation's death behold
> As Adam saw its prime!"

Then the last drops of ocean are dried up, and nothing but a bare and blighted mass of rock is left of that earth which once, even in its Maker's eyes, was altogether good. Still the doomed planet rushes on to its awful fate. Swiftly and more swiftly it circles round the sun, like the bark which once drawn within the influence of the whirlpool is sucked irresistibly into its fearful vortex. At last it seems to get paralyzed by the iron grasp that is tightening upon it—it staggers, pauses for a moment in its headlong career, and thus checked in its onward progress the sun draws it straight down to itself. A hurried rush through the tossing sea of fire, a swift plunge through the cloudy stratum behind, and the earth sinks to its eternal resting-place on the face of its parent globe.

THE STARS.

WE must turn now to realms lying beyond our own solar system. Beginning with ourselves and our sister planets, we have considered in succession the lesser light which rules the night, the greater light which rules the day, the comets which wing their wild flights around it, and the mysterious ether which encompasses it, and which in time will bring all the rest—planets, satellites, and comets, within its remorseless grasp. And now it remains to wind up by taking a glance at the great stellar universe around us, compared with which our own system, mighty as it is, counts but as a drop in the ocean. And here, as formerly, our subject naturally divides itself into two parts. We have to consider first, those facts relating to the stars which depend upon their general physical character and their individual peculiarities; and secondly, those which depend on their collocation in space and their movements through the realms of the infinite. As for the first, the great distance at which the stars are from us prevents us from knowing almost anything whatever about their condition, except what we can infer from analogy. They hold the same place in creation that our own sun does. They are not satellites of any other body, but primary orbs, independent sources of light and heat, and probably the centres of systems not less varied and gorgeous than our own. Hence we may argue with a high degree of probability that those facts which have been ascertained concerning the general nature of the sun, hold equally true of the stars. And as for their individual peculiarities, we are for the most part equally in the dark about them also, and that for the same reason. All the stars appear to us as mere luminous specks without any perceptible magnitude. And although "one star differeth from another star in glory," though even the naked eye can detect many degrees of brilliancy among them, yet all we can infer from this is that the more brilliant ones are probably much nearer to us than the others. But there are two classes of stars which form marked exceptions to the general rule, and stand out prominently from the rest. These are binary stars, and periodical or temporary stars.

The existence of the binary stars was discovered by Sir William Herschel towards the close of last

I

century. It had long been noticed by astronomers as a remarkable coincidence that in several instances a pair of bright stars were found in very close proximity to each other, much closer than we should have expected supposing the stars to have been scattered up and down at random over the whole face of the heavens. Still it was never thought that this was anything more than a coincidence; it was supposed that the stars had no connection with each other, but were altogether separate bodies, which merely happened to be situated in one straight line with ourselves. But Herschel having, for some scientific purposes which it would take too long to explain, determined to make a series of minute and careful observations upon these double stars, soon found to his surprise that they were rapidly shifting their positions relatively to each other; and, in short, he was ere long led to the conclusion that the two stars were in reality situated close together, and revolving in orbits round one another. Many pairs of stars of this kind were observed and registered, while in some cases the combinations were found to consist of three stars, and even four, instead of two.

But one of the most remarkable features about these multiple stars is that they are very frequently of different colours. In the case of the double stars the two colours are usually complementary; colours, that is, which when mixed together, in proper proportions, produce white. Thus one will be green and the other red, or one orange and the other blue, or one violet and the other yellow. Similarly in the triple stars we may have a blue, a red, and a yellow, or a green, an orange, and a violet. In a quadruple star we may have blue, green, orange, and red; and so on, in endless combinations. If there be any planets in attendance upon these multiple suns, as in all probability there will be, the celestial phenomena at those planets will be of the most extraordinary character indeed, and everything that depends on these phenomena—their times and their seasons, their days and their years—will be involved in the most intricate complications. If indeed any of them happened to be situated in very close proximity to one of the primaries, things with it would not be so confused. It would always revolve round the same sun, though in a very irregular and perturbed orbit; and hence its days and its years would follow each other pretty much in the natural and regular order. But its seasons will vary much both in length and temperature, and its nights, though much darker than its days, will yet differ from them far less than is the case with us. For when the primary orb sinks beneath the horizon, the secondary ones will shine out in full splendour, much smaller and more distant than the primary, but yet far exceeding in brilliancy the borrowed light of the brightest of full moons. But most of the planets, not nestled close enough beside any one of their suns, will come pretty equally under the influence of all. Take for instance the case of a planet in a quadruple system at a time when it happens to be about equally distant from all its four suns. A green and a red sun are above the horizon, and when we look directly at either, its colour is clear, brilliant, and well-defined. But their rays meet and mingle and unite into a dazzling snowy white, which imparts to the whole landscape the pure radiant look which seems to fill the firmament on a sunny day when the ground is covered with snow. A light cloud-wreath steals over the green sun, and a faint rosy blush overspreads the face of the sky. The cloud thickens and the rosy hue deepens into a mellow crimson. Then the green sun sets and a blue one rises, changing the red light of the sky into a rich purple, veined here and there with pale amethyst, as a few rays from the green sun struggle through the clouds just as it sinks beneath the horizon. The purple changes into a deep gold as the blue sun is succeeded by an orange one, and the gold pales down as the red sun sinks to his rest in turn. The orange is left alone, and when it, too, sets, night comes on apace. And now the moons rise and shed their radiance on the scene. But how differently do they show from the pale uniform light that beams from our own plain satellite! Every colour of the rainbow glows from their faces; in belts, in spots, in lunes, their chequered discs reflect every shade of hue that the artist's palette can produce. The parts illumined by one sun alone reflect, more faintly than the rest, the colours of their respective orbs; those which come within the light of two or three of them will shine more brightly and with gayer combinations of colours; while in the parts on which all the four suns shine at once we find again the snowy white, so bright as to sparkle almost with the light of day. But where there are four great lights to rule the day, night will be of unfrequent occurrence and of short duration; and soon the four suns, their nocturnal course ended, begin at once to draw nigh to their rising. Pale, slender threads of red, green, blue, and orange steal out from the darkness in four quarters of the horizon; and these widen and lengthen till they mingle together at their extremities in softly shading hues of white, indigo, and gold. Brighter and broader they grow, and the gorgeous variegated belt spreads rapidly from horizon to zenith, till at last the suns have fairly risen, and their many-coloured rays combine again into the dazzling white of the perfect day.

Nor are the annual phenomena of these planets—those, namely, which are connected with their seasons and their years—less extraordinary than their diurnal phenomena, which relate to their days and nights. Take the case of a planet in circumstances such as we have supposed above, situated at about equal distances from its several suns. It has just returned from performing a revolution about one of them, and while away on the farther side of that body it was pretty far removed from the influence of the others; and hence it has enjoyed a tolerably quiet and orderly year. Its days and its seasons have followed each other in due succession, though in their length and their temperature there were many and varied irregularities. But now, completing its circuit, it comes again into the region of confusion and anarchy. New suns wax and brighten till they rival the old one in splendour. Distinction of day and night there is none. Universal summer prevails over the planet—in some places mild, in others extreme—these patches of different climates being seemingly scattered up and down, arbitrarily over the face of the globe.. But gradually one of the suns—not the same one as before—begins to exercise a markedly more potent influence than the rest; and they slowly dwindle in the distance, while the victorious orb grows larger and brighter as it draws its captive down, towards it. And now the planet starts to perform; again a new revolution in a comparatively undisturbed orbit. Day and night, summer and winter, seed-time and harvest resume their wonted sway. But how altered are they all! The sun and the moon have changed their size, their brilliancy, and their colour; new planets stud the sky; new comets wheel around the sun; and only the more distant stars retain their positions unaltered. The year has changed its length; seasons and climate are revolutionized; zones formerly frigid or tropical have become temperate; or those temperate, torrid. The old vegetation, blighted by drought or nipped by unwonted frost, withers away, and new trees and plants take its place. The fauna change with the flora; birds and animals migrate; while whole races of men follow their example, or adapt themselves with difficulty to their altered climates. But scarcely are they settled down to their new circumstances ere a similar change again takes place, and they are whirled off to perform an equally brief circuit around yet another sun. And so on they go in their restless career—

"It may be for years, and it may be for ever,"
unless a favouring chance carries their planet very near to one of its suns, and thus enables that orb to establish an indisputable sway over it, and secure it as a permanent satellite.

The second class of stars to which we alluded—the periodical and temporary stars—are much rarer phenomena than those which we have just been considering, while their actual nature is altogether unknown to us. There are on record about six or eight instances in which bright stars have suddenly appeared in parts of the heaven where previously there was none, and after continuing to shine with varying brilliancy for a few months or years, have been again utterly extinguished. In one case, at least, it is believed that a star of this kind has reappeared several times at intervals of about one hundred and fifty years—a fact which, if true, would seem to indicate some periodicity in the causes of its appearances and disappearances. In several cases of a somewhat similar kind this periodicity is unquestionably to be found. A good many stars undergo regular increases and diminutions of brilliancy, the period of some being two or three days, of others as many months, and of others several years. Some of them, at stated intervals, disappear for a short time altogether. Various theories have been suggested to account for these phenomena. Some astronomers suppose that the stars in question have large dark planets or companion suns revolving round them, which at intervals interpose between them and us, and cut off the whole or a part of their light. Others attribute their varying brilliancy to dark spots and patches upon their surfaces. The temporary stars some suppose to have been altogether annihilated, or to have had their sources of light and heat exhausted. Others suppose them to revolve in very long orbits like the comets, and only to become visible when at the parts of their course nearest to the earth. All these hypotheses are possible, but hypotheses it is to be feared they will ever remain. Nothing but actual observation could tell us which is the correct one, and the bodies in question are so distant that evidence of this kind we can never hope to obtain.

Of the physical constitution of the stars we know but little. Analogy tells us that they are bodies of the same character, and probably of much the same magnitude, as our own sun. Recent spectrum analysis goes further, and shows us from an examination of their light that the substances which exist most plentifully in the sun's atmosphere, such as sodium, are also to be met with largely among the stars. More information about them than this we have not much hope of attaining to. There is no reasonable probability of our ever having telescopes powerful enough to give us further revelations of the nature of the stars. To our present instruments they appear simply as specks of light of no visible dimensions, and differing only in brightness. According to these varying degrees of brilliancy the stars are classed—the brightest being styled of the first magnitude, the next of the second magnitude, and so on through the telescopic stars down to the fifteenth. But this term must not be misunderstood. None of them have any perceptible magnitude whatever; even Sirius, the brightest, presents no marked disc like the planets; he is strictly a mathematical point of light—position without magnitude. It is probable that the stars do not differ very much in actual size and inherent brilliancy, and that their gradations of apparent brightness are due almost entirely to the different distances at which they are situated from us. The tiny orb, which is only revealed to us by the most powerful telescopes, is probably a not less glorious sun than Sirius or Procyon, but it is buried at such a depth in the abysses of space as to be altogether invisible to the unaided eye.

Sir William Herschel was the first astronomer who compared actually the light emanating from the different stars, and calculated from this their probable relative distances. This was done without much difficulty, and he next set himself the great task of discovering from these data the way in which the stars are distributed through space, the configuration of the great stellar universe, and the position which our sun occupies in it. The labour required for this was immense. It involved a careful examination of every part of the sky, both in the northern and southern hemispheres, and a tabling of the number of stars in them arranged in order of magnitude. He patiently directed his telescope by turns to the different quarters of the heavens, and calculated first the number of the larger stars which were to be found in them. The result of this showed that the first three or four magnitudes were distributed about equally over the whole sky; and he accordingly inferred that to a certain distance at least, the stars were grouped uniformly all round us. But then came a sudden break. When he counted those next in order, he found that except in one portion of the sky there were scarcely any; and when he arrived at the telescopic stars—those namely which are invisible to the naked eye—there were none at all, except in that great luminous belt which spans the whole firmament, and which is known by the name of the Milky Way. The inference from these facts was obvious. In most directions the stars came very speedily to an end, but in one circle all round us they seem to extend almost to infinity. We all know the appearance presented to the naked eye by the Milky Way—a white fleecy background, dotted all over with bright and distinct stars. When the telescope is turned upon it, we get a step further; some of the white background is in its turn resolved into separate stars, but another indistinct layer rises up behind them in turn. Higher and higher telescopic powers were attained, but still with the same result; the astronomer's eye penetrated further and further into the depths of space, but still the dim white background of star-dust filled the field of the glass. At length, however, by some optical improvements of Herschel's own devising, he succeeded in considerably increasing the efficiency of his telescopes, and was rewarded by finding the last layer of star-dust completely resolved into distinct and separate orbs. Background there was now none;

clear and bright the last stars shone out from the deep black void of the midnight sky. The astronomer laid down his glass; the furthest limits of our universe had been sounded, and its bounds assigned it in every direction. Laying down the telescope, Herschel took up the compasses, and proceeded to map on paper the results of his long and patient search. The conclusion he arrived at was, that the starry universe formed a roundish but irregular disc, with a deep cleft at one side extending nearly down to the sun, which occupies pretty much the centre of the disc. Thus viewed from above or below, the appearance presented would be circular, while laterally it would be that figured in the diagram. *S* is the

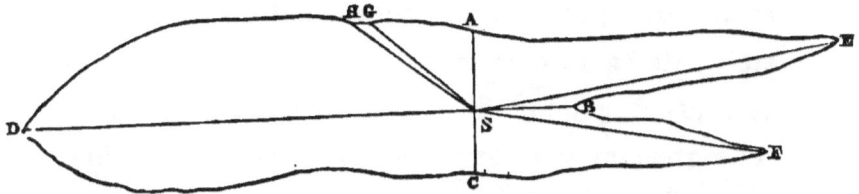

sun, and if we look in any of the directions *SA*, *SB*, *SC*, the range of stars which we see is a very brief one; but looking along *SD*, or any other direction in a plane through it perpendicular to that of the paper, we have a much farther vista. These directions accordingly correspond to the Milky Way in the heavens. In one part the Milky Way splits up into two branches, separated by a short interval from each other. These two branches are marked in the figure by the lines *SE*, *SF*. The constellations are wedges cut out of this great star-disc. Thus the Great Bear, for example, lies between the lines *SG*, *SH*, and therefore contains no very small stars. The Swan, on the other hand, lies along *SD*, and is hence enriched by a brilliant back-ground of star-dust. Herschel calculated that the number of stars in this enormous cluster is certainly not less than five-and-a-half millions, and is probably one or two millions more.

But a harder problem still remained behind. The relative distances of the stars had been computed, the farthest bounds of the known universe had been pierced, its figure determined, and its limits assigned it in every direction. But of the absolute dimensions of this great system, astronomers in Herschel's time knew nothing. They had learnt that the farthest stars were about five hundred times more distant from us than the nearest ones. But the distance of even the nearest was a sealed mystery. It was known that it could be measured only by millions of millions of miles; but how many of these great units it contained they could not tell. The way in which the distance of the nearer heavenly bodies is found is very simple.

When a land surveyor wishes to ascertain the distance of an inaccessible object, he measures what is called a base-line; and making an observation first from one end of this and then from the other, the displacement in the apparent position of the object gives him its distance from either station. And this is the method pursued with the sun and moon. They are observed simultaneously from two distant points of the earth's surface, and from their consequent displacement—or parallax, as it is called in scientific parlance—their distance is readily found. But no such plan avails us with the fixed stars. The distance of any two stations on the surface of our globe is as nothing compared with the enormous space which separates us from them. But the earth's motion in its orbit comes to our aid, and gives us two stand-points immeasurably farther apart from each other than its own two poles. Let us carefully observe one of the stars now, and then wait till six months have elapsed. The earth's revolution about the sun has brought us to a station two hundred million miles away from that of our first observation. And here surely with such an enormous base-line for our observations the difficulty will be easily solved—the star will appear to have shifted to quite a new part of the heavens. But no—to an ordinary telescope no change whatever is visible. One of the chief arguments of the old astronomers against the Copernican system was founded upon this; if the earth revolved in an orbit of such magnitude, the stars should appear in altogether different positions at different seasons of the year. Copernicus, with a little hesitation, gave the true reply—the stars must be at such a distance from us that our orbit is but as a speck in comparison with it. The astronomers laughed this to scorn; it was impossible to believe that the works of creation should have such vast extent as this explanation would involve. But they were attempting to limit the power of the Almighty. In the eternity of God a thousand years appear as one day; to His infinitude a million million miles are but a single span. Even the distance they thought so impossibly great, we, with the new light of science, know to be but an infinitesimal fraction of the whole dimensions of the universe.

From the days of Copernicus down to the present time, telescopes have steadily been increasing in power and efficiency. Astronomers were well aware that the problem of finding the stellar parallax was only an instrumental difficulty; the stars must suffer some displacement according to the season of the year, and only a sufficiently good telescope in the hands of a sufficiently skilful and accurate observer was required to determine it. By comparing the stellar light with the solar, they had arrived at a rough idea of the probable distance and parallax of some of the stars, and had found reason to think that the parallactic angle was not so small as to make its detection a hopeless task. Several generations of astronomers, however, passed away, and left this, one of the great objects of their life, unfulfilled. The great glory of discovering stellar parallax is shared equally by a Scotch and a Prussian astronomer, Henderson of Edinburgh and Bessel of Köningsberg, who made at the same time, about thirty years ago, independent observations of two different stars. The extreme difficulty of the problem will be seen when we state that the displacement found by Bessel, that of a star in the neck of the Swan, is only an angle of a third of a second, or less than a five-thousandth part of the apparent diameter of the sun. We must remember that every astronomical observation is subject to a host of errors. The most perfect telescope ever set up has a score of imperfections in construction and adjustment, and these have to be carefully calculated from repeated observations, and allowed for by difficult mathematical processes. There are also many astronomical sources of error, such as the finite velocity with which light is propagated, and the refraction of its rays by the earth's atmosphere. All these causes produce a far greater displacement in the star than its actual parallax, and the successful elimination of them all, and determination of the small residual angle, is justly regarded as one of the greatest triumphs of human skill and ingenuity. The parallax of these stars having been found, their distances were given at once by a simple trigonometrical formula. The result is that the nearest star is two hundred million times farther from the sun than the earth is,—in other words, the distance of the sun from its nearest neighbour in the great stellar cluster is twenty millions of millions of miles. It follows that the full dimensions of that great cluster from end to end must be at least twenty thousand billion miles.

One problem more, the last and the loftiest, yet remains to be solved, before we can say that we have completely mastered the system of the universe. We have calculated the dimensions of the great stellar cluster, we have determined its configuration, we have estimated the number of orbs which it contains. The question yet remains whether the stars are really fixed, as their popular name supposes, or whether they, like all the minor bodies we have considered, have their own special orbits and revolutions. Satellites circle round their primaries, planets wheel in obedience to the behest of their parent suns, comets under the same potent spell wing their fiery flight through space. And are there no fixed centres amid all this ceaseless motion, no spots on which the wearied imagination may settle, and contemplate from a solid and stable standpoint the workings of the great mechanism around? Science answers, there is none. Wherever there is matter there must be gravitation. The greatest and most glorious orbs of heaven are not less fully bound by that all-pervading law than the lightest speck of sea-foam or the filmy texture of the comets. The sun himself, upon whose majestic court hundreds of bright attendants wait, is subjected in turn to the influence of his mighty brethren, and rolls at their bidding along his appointed course.

We have seen that the motion of the earth in its orbit causes a displacement in the apparent positions of the stars. This displacement however is only a periodical and temporary one; as the seasons circle round, the earth returns to the spot from whence he set out, and the stars resume their old positions. The fact that some of them had a distinct and separate motion, indicating a permanent change of their position relatively to the sun, was first discovered by Edmund Halley. Some observations of the three brilliant stars, Sirius, Arcturus, and Aldebaran, made by the old Egyptian astronomers, had fortunately been handed down to his time, and on looking over them, he perceived that these stars must have shifted their positions since that early time, by a small but well-marked amount. This indicated that either these stars, or the sun, or probably both, must have changed their places by many million miles since those old records were penned by the philosophers of Alexandria. Other astronomers followed in Halley's track, and by the beginning of this century the proper motions of more than a hundred stars had been determined, chiefly by comparing them with Tycho Brahe's catalogue, made out two hundred years before. These proper motions showed great differences in amount and in direction, and no attempt was made to reconcile and systematize them until the subject was taken up by the bold and speculative genius of Sir William Herschel, who revelled in difficulties, and whose daring and ambitious spirit always selected the loftiest and apparently most hopeless themes. He succeeded in evoking order out of apparent confusion and chaos, and announced his discovery of the fact, that the sun, with all his gorgeous following, is sweeping majestically through space in the direction of the constellation Hercules. It was not till fifty years afterwards that another astronomer was found bold enough to grapple with this mighty theme. It was then taken up by some of the leading astronomers of Russia, with the advantage of half a century's additional observations, and Herschel's results were confirmed in the fullest manner possible. The direction in which the earth is moving is now known beyond the possibility of a doubt. His velocity, however, has been variously estimated at from thirty million to a hundred million miles per annum.

Of course the other suns of our great cluster have their own motions also; their varying position relatively to ourselves depending partly upon our motion and partly on their own. Mathematical theory, proceeding upon Newton's great law, tells us that the centre of this universal motion must be the centre of gravity of the whole stellar cluster; that any star situated there must be at rest, while all the others are circling in ceaseless revolution around it. Mädler, of Dorpat, is the only astronomer who has ventured to seek for this central sun. By studying Herschel's diagram of the stellar system, and combining it with the known direction of our sun's motion, he was led to believe that the centre of gravity of that system must be situated in or near the constellation Taurus. A careful examination of all the stars in that quarter of the heavens made him finally fix upon Alcyone, the central orb of the Pleiades, as being the object of his search. It is probable that his speculations are somewhat premature; the data upon which they are founded are slight and partially uncertain, on account of the extreme slowness of the motions from which they are deduced. In fact it is probable that many generations must pass away before a sufficiently long course of observation can either fully confirm or disprove the conclusions at which he arrived. If his theory be correct, the sweet influences of the Pleiades must be potent influences indeed. Holding their eternal court unmoved in the centre of the heavens, they send out their resistless influence to the farthest confines of space, and bend into stately curves around them the most distant bodies of the universe, some of whose grand orbits cannot be traversed in less than five hundred million years.

THE NEBULÆ.

THE astronomer has now completed his investigation of the great stellar system of which our sun forms a member. The figure of that system has been defined, its dimensions calculated, and its motions traced out and analyzed. This was the consummation to which, until very recent periods, astronomers of all ages had been accustomed to look as the great goal to which their science tended, though a goal which even the most sanguine of them scarcely hoped that it would ever actually reach. And now that it has been so fully attained, can the astronomer sit down and rest on his laurels and boast that he has fathomed the remotest depths of the universe of God? Not so. While with slow and toilsome steps he has been creeping up to his first goal, he has at the same time been seeing another gradually emerging from the obscurity in front of him, and now that the former is reached, it is but to see the latter standing clearly out in the distance, more hopelessly inaccessible than the other had ever seemed to be. For when we stand on the remotest orb of the Milky Way the telescope reveals new marvels to our gaze, and opens up fresh and undreamt of regions for scientific research. Not yet are the wonders of creation exhausted. Had it been so, had these really been the uttermost bounds of God's created works, we should still have pronounced them well worthy in magnificence and grandeur of the Omnipotence which called them into being. Let us think for a moment on what a scale of inconceivable magnitude the universe thus far is built. The orbit of our earth is two hundred million miles in diameter, but so insignificant is this vast distance compared with the great gulf which separates us from the fixed stars, that to all except the very most powerful telescopes those stars seem to occupy exactly the same positions in the sky when viewed from two opposite points of our course.

Try a still farther flight,—pass over twenty millions of millions of miles. We have reached the nearest of the stars, and taking our stand on one of its planets, and waiting till evening falls, we look eagerly abroad to mark the altered aspect of the heavens. Here surely, where we have put such an overwhelming distance between us and our former position, the face of the sky will be no longer recognizable—the old heavens will have passed away from over our head, as well as the old earth from beneath our feet. But no,—as the stars one by one steal out from the darkness, they group themselves into their old well-known configurations. There is the Little Bear with its pole-star, and the Great Bear with its pointers, there are the bands of Orion and the sweet influences of the Pleiades, there are Mazzaroth and Arcturus, just as they appeared to Job five thousand years ago, and sixty billion miles away. Vast as is the space we have traversed, it is not a thousandth part of that which separates the two most distant stars of the system, and hence we need not wonder that the change we have found is no greater than that which comes over the distant landscape as the traveller advances a score or two of yards along his way. Let us then pursue our journey still further. Sun after sun beams upon us with its brilliant band of planets and comets,—sun after sun pales and lessens in the distance as we leave it behind in our flight. And gradually a change creeps over the face of the heavens. The general figures of the constellations remain the same, but those behind contract their dimensions and shrink more closely together, while those in front are opening out and growing larger and brighter. At length we near the farthest confines of the Milky Way. Very few and very scattered are the stars which still remain in front of us. We can number them all with ease. And now but three are left before us,—but two,—but one. That one is reached in turn. We pass to the further side and look forth into the mysterious abyss which lies beyond. Before, behind, to the right, to the left,—whichever way we turn our gaze, it meets with nought but the blackness of darkness,— the deep gloom of the midnight sky is unbroken by the gleam of a single star. Onward still we wing our daring flight; the last resting-place is abandoned, the last oasis left behind, and we adventure forth into the trackless wastes of space. One by one the planets of this last sun are passed in our course; now and then a comet overtakes us, and blazes swiftly past into the depths beyond; but if we look onwards, we see that even it soon pauses in its reckless flight, and wheels back on rapid wing to less solitary and untrodden regions. The sun itself dwindles down to a star, and takes its place among a cluster of others which come forth from behind and around it as its paling light permits them to become visible. And soon this cluster too fades, till all distinction of stars in it is lost, and nothing is left but a dim white patch of light, ere long to be blotted out in turn, as it seems to be swallowed up by the surrounding darkness. All created works are left behind, and we stand alone face to face with the infinitude of God,—alone where mortal footstep has never trod, where presence there has never been, save that of the ever omnipresent Creator and the spirits which pass and repass, ascending and descending the ladder of vision which bridges the chasm between heaven and earth, as they go and come, ministering to the heirs of salvation.

But suppose that our vision is now quickened by telescopic aid. We turn the glass in the direction from which we have just come to get a last look of the universe we are leaving behind us. And, surely enough the little white patch steals out again, from the darkness in the centre of the field of view. But what is that faint film of light that clouds the outer edge of the circle? Turn the full power of the instrument upon it, and it brightens into a white patch exactly similar to the former. Move the telescope round, and another, and another, meets the eye. Direct it to any new quarter of the heavens, and the sight presented is the same. The whole sky is mottled with these flecks of white. Thick they are as the motes that dance in the sunbeam, close as the stars that stud the firmament, innumerable as the grains of sand upon the shore. In aspect they exactly resemble the dim and distant appearance presented to view by the system which we have just left,—so exactly, that when once the glass has been turned away from it, we find it impossible, on turning back, to pick it out from the multitudes that surround it. Let us choose out one of the nearest and brightest of these, and wing our flight towards it. As we approach it, we might think that we were but returning to the system we had left. The white patch widens and brightens, stars emerge from it by thousands and tens of thousands, till it fills the whole firmament with a blaze of splendour, surpassing by many fold the brilliancy of our own nocturnal sky. For the telescope teaches us that while each of these objects is a system of stars probably not inferior in glory to that of which we form a part, many of them must surpass it in magnificence at least a thousand-fold.

It would be hopeless to attempt expressing in ordinary language the vast distance at which these clusters of stars are situated from us. If we were to reckon it in miles, or even in millions of miles, figures would pile upon figures, till in their number all definite idea of their value was lost. We must choose another unit to measure these infinitudes of space— a unit compared with which the dimensions of our own solar system shrink into absolute nothingness. The velocity of light is such that it would flash fifteen times from pole to pole of our earth between two beats of the pendulum. It bridges the huge chasm that separates us from the sun in little more than eight minutes. But the light that shows us these faint star-clusters has been travelling with this frightful velocity for more than two million years since it left its distant source. We see them to-day in the fields of our telescopes, not as they are now, but as they were countless ages before the creation of man upon the earth. What they are now, who can tell? Recent spectrum analysis reveals many of them to us as consisting not of separate stars, like their companion nebulæ, but of purely gaseous matter. They appear as vast oceans of flaming gas, doubtless much resembling what we have described as the condition of the solar envelope, but hanging altogether isolated in space, while they spread their huge billows over countless millions of miles. In others we see the work of condensation already begun, as their central parts show traces of solid, or at least of liquid, matter. In fact they realize Laplace's magnificent conception of the original state of our own system,—

a void chaotic mass, containing in it the germs of all the stars that lighten our midnight sky. But from all this we can argue nothing of their present state. It may well be that their transformation has been effected ages since, and that they too have long ago split up into countless myriads of suns.

And what we were when their light started on its mission towards us, God alone knows. A fiery globe of molten rock—a thin cloud of vapour without form and void—or, it may be, not yet in existence at all, unseen and undreamt of save in the eternal counsels of the Creator. Not till the rays had well-nigh completed their flight were the progenitors of those for whose eyes they were destined first formed upon the earth. The six thousand years of man are as nothing compared with their long journey. Two hundred generations have come and gone since Adam walked in Paradise, but if each of these generations had been told over two hundred fold, the antiquity of man would not have rivalled the hoary age of these tiny waves of light. So small are the undulations of ether, that forty thousand of them follow each other within the space of a single inch, so impalpable that they beat upon our eyeballs in countless millions and with incalculable velocity without injuring their delicate surface; and yet so infinitely strong, when guided by their Maker's hand, that they have steered their unerring course for millions of billions of miles, and reached their destination unaltered and unenfeebled.

Of a truth "things small and great," the infinitesimal and the infinite, "bless the Lord; they praise Him and magnify Him for ever." What overwhelming force in these words of God when read by the light of His works:—"When I consider Thy heavens, the work of Thy fingers, the moon and the stars which Thou hast ordained; what is man that Thou art mindful of him, and the son of man that Thou visitest him?" And what unspeakable preciousness in these:— "Thus saith the Lord, the heaven is my throne, and the earth is my footstool; but to this man will I look, even to him that is poor and of a contrite spirit and that trembleth at my word."

And now let us go one step further, and ask ourselves where these magnificent creations, rising above each other in the scale of magnitude and grandeur, are to have an end. First we had planets with their attendant satellites, then suns with their accompanying planets, then a great cluster consisting of many million suns, and now this vast system composed of thousands of similar clusters. How many steps more we might go in this ascending scale before we reached the climax, it is impossible to say. The system of clusters which we have above described may be itself but a unit in a yet greater and more gorgeous whole, and that whole in turn an insignificant fraction of a creation grander yet. But wherever the end may be, to us in our present state it has already come. Further than we have now reached, our finite powers of observation can never hope to attain. A veil of impenetrable obscurity is drawn over all that lies beyond. The Almighty has interposed His stern fiat before the advancing flood of human science, which would fain overspread all creation with its triumphant billows. "Hitherto shalt thou come, but no further, and here shall thy proud waves be stayed." The glories that lie beyond are among those things which eye hath not seen, which ear hath not heard, and which it hath not entered into the heart of man to conceive. But though we cannot trace the steps in the ascending scale of creation, we may soar past them in imagination, and ask what the great climax will be. God himself is infinite, but His works must be finite. Suns, clusters, and systems may rise above each other in almost endless succession, but at length some one great system must be reached, whose members, circling round and round in harmony among themselves, must include within their vast limits all the works of the Creator. And what is the mighty centre around which all these motions take place, the one fixed spot in the universe about which all else is in rapid and ceaseless revolution? What can be the worthy centre of this magnificence, the glory compared with which all others sink into the shade? There is indeed a glory that excelleth, a glory such that in the apostle's words even "that which was made glorious hath no glory by reason of the glory that excelleth." But this is a glory uncreate, a glory that shines with no borrowed splendour; it is the glory of the Godhead. Men speak much of heaven as consisting in the felt nearness and presence of God, and hence of its being a state rather than a place. But they forget that while God himself is infinite, the human nature of our Lord and Saviour cannot be so, and hence that there must be some one spot dignified above all others by a special Shechinah—a special manifestation of the glory of the Godhead—that spot where the seer of Patmos beheld a throne set in heaven, and One that sat upon the throne, and in the midst of it a Lamb as it had been slain. And if we ask again what can be the fitting centre of all the gorgeous systems which the science of astronomy reveals to our astonished gaze, what answer *shall* we, what answer *can* we, give but one? Surely we must say with the patriarch, "THIS IS NONE OTHER THAN THE HOUSE OF GOD, THIS IS THE GATE OF HEAVEN."

THE END.

W. METCALFE AND SONS, PRINTERS, CAMBRIDGE.

New Edition, 18mo., *Price* 4/6.

POPULAR ASTRONOMY.

Lectures by

SIR G. B. AIRY, K.C.B.,
ASTRONOMER ROYAL.

WITH NUMEROUS ILLUSTRATIONS.

The speciality of this volume is the direct reference of every step to the Observatory, and the full description of the methods and instruments of observation. *The Guardian* says—"Popular Astronomy in general has many manuals, but none of them supersede these lectures."

New Edition, 18mo., *Price* 5/6.

ELEMENTARY LESSONS IN ASTRONOMY.

By J. NORMAN LOCKYER, F.R.S.

With Coloured Diagram of the Spectre of the Sun, Stars, and Nebulæ, and numerous Illustrations.

"The book is full, clear, sound, and worthy of attention, not only as a popular exposition, but as a scientific index."—*Athenæum.*
"The most fascinating of elementary books on the sciences."—*Nonconformist.*

MACMILLAN AND CO., LONDON.

Second Edition, in Imperial 8vo. cloth, extra gilt, Price 31/6.

Illustrated by Eleven Coloured Plates and 455 Woodcuts.

THE FORCES OF NATURE.

A Popular Introduction to the Study of Physical Phenomena.

By AMÉDÉE GUILLEMIN.

Translated from the French by Mrs. NORMAN LOCKYER, and Edited, with Additions and Notes, by

J. NORMAN LOCKYER, F.R.S.

Contents :

Book I. Gravity.—Book II. Sound.—Book III. Light.—Book IV. Heat.—Book V. Magnetism.—Book VI. The Electric Light.—Book VII. Atmospheric Meteors.

"This book is a luxurious introduction to the study of the physical sciences. M. Amédée Guillemin, in his splendid work on 'The Heavens,' has popularized some of the greatest discoveries of astronomy, and the present work will do the same for physics. The method of pictorial illustration, accompanied as it is by descriptions of singular clearness, makes the experiments as easy to understand as though they were actually performed before the reader. There are 455 of these woodcut illustrations, all well executed, and so admirably fitted to the text as to make the book interesting to young people, while it is at the same time worthy of the notice of the student. M. Guillemin has found an excellent translator in Mrs. Norman Lockyer, while the editorship of Mr. Norman Lockyer, with his notes and additions, are guarantees, not only of scientific accuracy, but of the completeness and lateness of information."—*Daily News.*

"Altogether the work may be said to have no parallel, either in point of fulness or attraction, as a popular manual of physical science."—*Saturday Review.*

Third Edition, revised throughout, with all the most recent Discoveries. Royal 8vo., cloth extra, Price, 21/-

SPECTRUM ANALYSIS.

By H. E. ROSCOE, F.R.S.,

PROFESSOR OF CHEMISTRY IN OWENS' COLLEGE, MANCHESTER.

Six Lectures with Appendices, Engravings, Maps, and Chromolithographs.

The Westminster Review says, "The Lectures themselves furnish a most admirable elementary treatise on the subject, which, by the insertion in appendices to each lecture of extracts from the most important published memoirs, the author has rendered it equally valuable as a text book for advanced students."

MACMILLAN AND CO., LONDON.

BEDFORD STREET, COVENT GARDEN, LONDON,
November, 1872.

MACMILLAN & Co.'s CATALOGUE of Works in the Departments of History, Biography, and Travels; Politics, Political and Social Economy, Law, etc.; and Works connected with Language. With some short Account or Critical Notice concerning each Book.

HISTORY, BIOGRAPHY, and TRAVELS.

Baker (Sir Samuel W.)—Works by Sir SAMUEL BAKER, M.A., F.R.G.S. :—

THE ALBERT N'YANZA Great Basin of the Nile, and Exploration of the Nile Sources. New and Cheaper Edition. Maps and Illustrations. Crown 8vo. 6s.

"*Bruce won the source of the Blue Nile; Speke and Grant won the Victoria source of the great White Nile; and I have been permitted to succeed in completing the Nile Sources by the discovery of the great reservoir of the equatorial waters, the Albert N'yanza, from which the river issues as the entire White Nile.*"—PREFACE. "*As a Macaulay arose among the historians,*" says the READER, "*so a Baker has arisen among the explorers.*" "*Charmingly written;*" says the SPECTATOR, "*full, as might be expected, of incident, and free from that wearisome reiteration of useless facts which is the drawback to almost all books of African travel.*"

THE NILE TRIBUTARIES OF ABYSSINIA, and the Sword Hunters of the Hamran Arabs. With Maps and Illustrations. Fourth and Cheaper Edition. Crown 8vo. 6s.

Sir Samuel Baker here describes twelve months' exploration, during which he examined the rivers that are tributary to the Nile from Abyssinia, including the Atbara, Settite, Royan, Salaam, Angrab, Rahad, Dinder, and the Blue Nile. The interest attached to these portions of Africa differs entirely from that of the White Nile regions, as the whole of Upper Egypt and Abyssinia is capable of development, and is inhabited by races having some degree of civilization; while Central Africa is peopled by a race of savages, whose future is more problematical. The TIMES says: "It solves finally a geographical riddle which hitherto had been extremely perplexing, and it adds much to our information respecting Egyptian Abyssinia and the different races that spread over it. It contains, moreover, some notable instances of English daring and enterprising skill; it abounds in animated tales of exploits dear to the heart of the British sportsman; and it will attract even the least studious reader, as the author tells a story well, and can describe nature with uncommon power."

Barante (M. De).—*See* GUIZOT.

Baring-Gould (Rev. S., M.A.)—LEGENDS OF OLD TESTAMENT CHARACTERS, from the Talmud and other sources. By the Rev. S. BARING-GOULD, M.A. Author of "Curious Myths of the Middle Ages," "The Origin and Development of Religious Belief," "In Exitu Israel," &c. In Two Vols. Crown 8vo. 16s. Vol. I. Adam to Abraham. Vol. II. Melchizedek to Zechariah.

Mr. Baring-Gould's previous contributions to the History of Mythology and the formation of a science of comparative religion are admitted to be of high importance; the present work, it is believed, will be found to be of equal value. He has collected from the Talmud and other sources, Jewish and Mohammedan, a large number of curious and interesting legends concerning the principal characters of the Old Testament, comparing these frequently with similar legends current among many of the peoples, savage and civilized, all over the world. "These volumes contain much that is very strange, and, to the ordinary English reader, very novel."—DAILY NEWS.

Barker (Lady).—*See also* BELLES LETTRES CATALOGUE.

STATION LIFE IN NEW ZEALAND. By LADY BARKER. Second and Cheaper Edition. Globe 8vo. 3s. 6d.

These letters are the exact account of a lady's experience of the brighter and less practical side of colonization. They record the expeditions, adventures, and emergencies diversifying the daily life of the wife of a New Zealand sheep-farmer; and, as each was written while the novelty and excitement of the scenes it describes were fresh upon her, they may succeed in giving here in England an adequate impression of the delight and freedom of an existence so far removed from our own highly-wrought civilization. "We have never read a more truthful or a pleasanter little book."—
ATHENÆUM.

Bernard, St.—*See* MORISON.

Blanford (W. T.)—GEOLOGY AND ZOOLOGY OF ABYSSINIA. By W. T. BLANFORD. 8vo. 21*s*.

This work contains an account of the Geological and Zoological Observations made by the author in Abyssinia, when accompanying the British Army on its march to Magdala and back in 1868, *and during a short journey in Northern Abyssinia, after the departure of the troops. Part I. Personal Narrative; Part II. Geology; Part III. Zoology. With Coloured Illustrations and Geological Map.* "*The result of his labours,*" *the* ACADEMY *says,* "*is an important contribution to the natural history of the country.*"

Bryce.—THE HOLY ROMAN EMPIRE. By JAMES BRYCE, D.C.L., Regius Professor of Civil Law, Oxford. New and Revised Edition. Crown 8vo. 7*s*. 6*d*.

The object of this treatise is not so much to give a narrative history of the countries included in the Romano-Germanic Empire—Italy during the Middle Ages, Germany from the ninth century to the nineteenth—as to describe the Holy Empire itself as an institution or system, the wonderful offspring of a body of beliefs and traditions which have almost wholly passed away from the world. To make such a description intelligible it has appeared best to give the book the form rather of a narrative than of a dissertation; and to combine with an exposition of what may be called the theory of the Empire an outline of the political history of Germany, as well as some notice of the affairs of mediæval Italy. Nothing else so directly linked the old world to the new as the Roman Empire, which exercised over the minds of men an influence such as its material strength could never have commanded. It is of this influence, and the causes that gave it power, that the present work is designed to treat. "*It exactly supplies a want: it affords a key* to much which men read of in their books as isolated facts, but of which they have hitherto had no connected exposition set before them. We know of no writer who has so thoroughly grasped the real nature of the mediæval Empire, and its relations alike to earlier and to later times."—SATURDAY REVIEW.

Burke (Edmund).—*See* MORLEY (JOHN).

Cameos from English History.—*See* YONGE (MISS).

Chatterton.—*See* WILSON (DANIEL).

Cooper.—ATHENÆ CANTABRIGIENSES. By CHARLES HENRY COOPER, F.S.A., and THOMPSON COOPER, F.S.A. Vol. I. 8vo., 1500—85, 18s.; Vol. II., 1586—1609, 18s.

This elaborate work, which is dedicated by permission to Lord Macaulay, contains lives of the eminent men sent forth by Cambridge, after the fashion of Anthony à Wood, in his famous " Athenæ Oxonienses."

Cox (G. V., M.A.)—RECOLLECTIONS OF OXFORD. By G. V. Cox, M.A., New College, late Esquire Bedel and Coroner in the University of Oxford. *Cheaper Edition.* Crown 8vo. 6s.

"*An amusing farrago of anecdote, and will pleasantly recall in many a country parsonage the memory of youthful days.*"—TIMES. "*Those who wish to make acquaintance with the Oxford of their grandfathers, and to keep up the intercourse with Alma Mater during their father's time, even to the latest novelties in fashion or learning of the present day, will do well to procure this pleasant, unpretending little volume.*"—ATLAS.

"Daily News."—THE DAILY NEWS CORRESPONDENCE of the War between Germany and France, 1870—1. Edited with Notes and Comments. New Edition. Complete in One Volume. With Maps and Plans. Crown 8vo. 6s.

This Correspondence has been translated into German. In a Preface the Editor says:—

"*Among the various pictures, recitals, and descriptions which have appeared, both of our gloriously ended national war as a whole, and of its several episodes, we think that in laying before the German public, through* a translation, the following War Letters which appeared first in the DAILY NEWS, and were afterwards published collectively, we are offering them a picture of the events of the war of a quite peculiar character. Their communications have the advantage of being at once entertaining and instructive, free from every romantic embellishment, and nevertheless written in a vein intelligible and not fatiguing to the general reader. The writers linger over events, and do not disdain to surround the great and heroic war-pictures with arabesques, gay and grave, taken from camp-life and the life of the inhabitants of the occupied territory. A feature which distinguishes these Letters from all other delineations of the war is that they do not proceed from a single pen, but were written from the camps of both belligerents." "These notes and comments," according to the SATURDAY REVIEW, "are in reality a very well executed and continuous history."

Dilke.—GREATER BRITAIN. A Record of Travel in English-speaking Countries during 1866-7. (America, Australia, India.) By Sir CHARLES WENTWORTH DILKE, M.P. Fifth Edition. Crown 8vo. 6s.

"*Mr. Dilke*," says the SATURDAY REVIEW, "*has written a book which is probably as well worth reading as any book of the same aims and character that ever was written. Its merits are that it is written in a lively and agreeable style, that it implies a great deal of physical pluck, that no page of it fails to show an acute and highly intelligent observer, that it stimulates the imagination as well as the judgment of the reader, and that it is on perhaps the most interesting subject that can attract an Englishman who cares about his country.*" "*Many of the subjects discussed in these pages,*" says the DAILY NEWS, "*are of the widest interest, and such as no man who cares for the future of his race and of the world can afford to treat with indifference.*"

Dürer (Albrecht).—HISTORY OF THE LIFE OF ALBRECHT DÜRER, of Nürnberg. With a Translation of his Letters and Journal, and some account of his Works. By Mrs. CHARLES HEATON. Royal 8vo. bevelled boards, extra gilt. 31s. 6d.

This work contains about Thirty Illustrations, ten of which are productions by the Autotype (carbon) process, and are printed in permanent tints by Messrs. Cundall and Fleming, under licence from the Autotype Company, Limited; the rest are Photographs and Woodcuts.

Elliott.—LIFE OF HENRY VENN ELLIOTT, of Brighton. By JOSIAH BATEMAN, M.A., Author of "Life of Daniel Wilson, Bishop of Calcutta," &c. With Portrait, engraved by JEENS; and an Appendix containing a short sketch of the life of the Rev. Julius Elliott (who met with accidental death while ascending the Schreckhorn in July 1869.) Extra fcap. 8vo. 6s. Third and Cheaper Edition, with Appendix.

"*A very charming piece of religious biography; no one can read it without both pleasure and profit.*"—BRITISH QUARTERLY REVIEW.

European History, Narrated in a Series of Historical Selections from the best Authorities. Edited and arranged by E. M. SEWELL and C. M. YONGE. First Series, crown 8vo. 6s.; Second Series, 1088–1228, crown 8vo. 6s.

When young children have acquired the outlines of history from abridgments and catechisms, and it becomes desirable to give a more enlarged view of the subject, in order to render it really useful and interesting, a difficulty often arises as to the choice of books. Two courses are open, either to take a general and consequently dry history of facts, such as Russell's Modern Europe, or to choose some work treating of a particular period or subject, such as the works of Macaulay and Froude. The former course usually renders history uninteresting; the latter is unsatisfactory, because it is not sufficiently comprehensive. To remedy this difficulty, selections, continuous and chronological, have in the present volume been taken from the larger works of Freeman, Milman, Palgrave, Lingard, Hume, and others, which may serve as distinct landmarks of historical reading. "*We know of scarcely anything,*" *says the* GUARDIAN, *of this volume,* "*which is so likely to raise to a higher level the average standard of English education.*"

Fairfax (Lord).—A LIFE OF THE GREAT LORD FAIRFAX, Commander-in-Chief of the Army of the Parliament of England. By CLEMENTS R. MARKHAM, F.S.A. With Portraits, Maps, Plans, and Illustrations. Demy 8vo. 16s.

No full Life of the great Parliamentary Commander has appeared; and it is here sought to produce one—based upon careful research in contemporary records and upon family and other documents. "*Highly useful to the careful student of the History of the Civil War. . . . Pro-*

bably as a military chronicle Mr. Markham's book is one of the most full and accurate that we possess about the Civil War."—FORTNIGHTLY REVIEW.

Faraday.—MICHAEL FARADAY. By J. H. GLADSTONE, Ph.D., F.R.S. Crown 8vo. 4s. 6d. Second Edition, with Portrait.

This Sketch of the Life, many-sided Character, and Work of Faraday is founded mainly upon the Author's own reminiscences of his friend, upon documents hitherto unpublished, and sketches of the philosopher which are less generally known, so that it may be regarded as almost entirely an addition to what has already been published on the same subject. The Sections are:—I. The Story of his Life. II. Study of his Character. III. Fruits of his Experience. IV. His Method of Writing. V. The Value of his Discoveries.—Supplementary Portraits. Appendices:—List of Honorary Fellowships, etc.

Field (E. W.)—EDWIN WILKINS FIELD. A Memorial Sketch. By THOMAS SADLER, Ph.D. With a Portrait. Crown 8vo. 4s. 6d.

Mr. Field was well known during his life-time not only as an eminent lawyer and a strenuous and successful advocate of law reform, but, both in England and America, as a man of wide and thorough culture, varied tastes, large-heartedness, and lofty aims. His sudden death was looked upon as a public loss, and it is expected that this brief Memoir will be acceptable to a large number besides the many friends at whose request it has been written.

Freeman.—Works by EDWARD A. FREEMAN, M.A., D.C.L.

"That special power over a subject which conscientious and patient research can only achieve, a strong grasp of facts, a true mastery over detail, with a clear and manly style—all these qualities join to make the Historian of the Conquest conspicuous in the intellectual arena."— ACADEMY.

HISTORY OF FEDERAL GOVERNMENT, from the Foundation of the Achaian League to the Disruption of the United States. Vol. I. General Introduction. History of the Greek Federations. 8vo. 21s.

Freeman (E. A.)—*continued.*

Mr. Freeman's aim, in this elaborate and valuable work, is not so much to discuss the abstract nature of Federal Government, as to exhibit its actual working in ages and countries widely removed from one another. Four Federal Commonwealths stand out, in four different ages of the world, as commanding above all others the attention of students of political history, viz. the Achaian League, the Swiss Cantons, the United Provinces, the United States. The first volume, besides containing a General Introduction, treats of the first of these. In writing this volume the author has endeavoured to combine a text which may be instructive and interesting to any thoughtful reader, whether specially learned or not, with notes which may satisfy the requirements of the most exacting scholar. "The task Mr. Freeman has undertaken," the SATURDAY REVIEW says, "*is one of great magnitude and importance. It is also a task of an almost entirely novel character. No other work professing to give the history of a political principle occurs to us, except the slight contributions to the history of representative government that is contained in a course of M. Guizot's lectures . . . . The history of the development of a principle is at least as important as the history of a dynasty, or of a race.*"

OLD ENGLISH HISTORY. With *Five Coloured Maps.* Second Edition. Extra fcap. 8vo., half-bound. 6s.

"*Its object,*" the Preface says, "*is to show that clear, accurate, and scientific views of history, or indeed of any subject, may be easily given to children from the very first. . . . I have throughout striven to connect the history of England with the general history of civilized Europe, and I have especially tried to make the book serve as an incentive to a more accurate study of historic geography.*" The rapid sale of the first edition and the universal approval with which the work has been received prove the correctness of the author's notions, and show that for such a book there was ample room. The work is suited not only for children, but will serve as an excellent text-book for older students, a clear and faithful summary of the history of the period for those who wish to revive their historical knowledge, and a book full of charms for the general reader. The work is preceded by a complete chronological Table, and appended is an exhaustive and useful Index. In the present edition the whole has been carefully revised, and such improvements as suggested themselves have been introduced. "*The book indeed is full of instruction and interest to students of all ages, and he must be a well-informed man indeed who will not rise from its perusal with clearer and more accurate ideas of a too much neglected portion of English history.*"—SPECTATOR.

Freeman (E. A.)—*continued.*

HISTORY OF THE CATHEDRAL CHURCH OF WELLS, as illustrating the History of the Cathedral Churches of the Old Foundation. Crown 8vo. 3*s.* 6*d.*

'*I have here,*" the author says, " *tried to treat the history of the Church of Wells as a contribution to the general history of the Church and Kingdom of England, and specially to the history of Cathedral Churches of the Old Foundation.* . . . *I wish to point out the general principles of the original founders as the model to which the Old Foundations should be brought back, and the New Foundations reformed after their pattern.*" " *The history assumes in Mr. Freeman's hands a significance, and, we may add, a practical value as suggestive of what a cathedral ought to be, which make it well worthy of mention.*"—SPECTATOR.

HISTORICAL ESSAYS. Second Edition. 8vo. 10*s.* 6*d.*

The principle on which these Essays have been chosen is that of selecting papers which refer to comparatively modern times, or, at least, to the existing states and nations of Europe. By a sort of accident a number of the pieces chosen have thrown themselves into something like a continuous series bearing on the historical causes of the great events of 1870—71. *Notes have been added whenever they seemed to be called for; and whenever he could gain in accuracy of statement or in force or clearness of expression, the author has freely changed, added to, or left out, what he originally wrote. To many of the Essays has been added a short note of the circumstances under which they were written. It is needless to say that any product of Mr. Freeman's pen is worthy of attentive perusal; and it is believed that the contents of this volume will throw light on several subjects of great historical importance and the widest interest. The following is a list of the subjects:—*1. The Mythical and Romantic Elements in Early English History; 2. The Continuity of English History; 3. The Relations between the Crowns of England and Scotland; 4. Saint Thomas of Canterbury and his Biographers; 5. The Reign of Edward the Third; 6. The Holy Roman Empire; 7. The Franks and the Gauls; 8. The Early Sieges of Paris; 9. Frederick the First, King of Italy; 10. The Emperor Frederick the Second; 11. Charles the Bold; 12. Presidential Government. " *He never touches a question without adding to our comprehension of it, without leaving the impression of an ample knowledge, a righteous purpose, a clear and powerful understanding.*"—SATURDAY REVIEW.

A Second Series of HISTORICAL ESSAYS in the Press.

Freeman (E. A.)—*continued.*

GENERAL SKETCH OF EUROPEAN HISTORY. Being Vol. I. of an Historical Course for Schools. Edited by E. A. FREEMAN, D.C.L. 18mo. cloth. 3*s.* 6*d.*

The present volume is meant to be introductory to the Historical Course for Schools. It is intended to give, as its name implies, a general sketch of the history of the civilized world, that is, of Europe, and of the lands which have drawn their civilization from Europe. Its object is to trace out the general relations of different periods and different countries to one another, without going minutely into the affairs of any particular country. This is an object of the first importance, for, without clear notions of general history, the history of particular countries can never be rightly understood. The narrative extends from the earliest movements of the Aryan peoples, down to the latest events both on the Eastern and Western Continents. The book consists of seventeen moderately sized chapters, each chapter being divided into a number of short numbered paragraphs, each with a title prefixed clearly indicative of the subject of the paragraph.

THE UNITY OF HISTORY. The "REDE" LECTURE delivered in the Senate House, before the University of Cambridge, on Friday, May 24th, 1872. Crown 8vo. 2*s.*

THE GROWTH OF THE ENGLISH CONSTITUTION FROM THE EARLIEST TIMES. Crown 8vo. 5*s.*

The three Chapters of which this work consists are an expansion of two Lectures delivered by Mr. Freeman; appended are copious notes, the whole book forming a graphic and interesting sketch of the history of the British Constitution, from an original point of view. The Author shows that the characteristic elements of the British Constitution are common to the whole of the Aryan nations. His "object has been to show that the earliest institutions of England and of other Teutonic lands are not mere matters of curious speculation, but matters closely connected with our present political being. I wish to show" he says, "that, in many things, our earliest institutions come more nearly home to us, and that they have more in common with our present political state, than the institutions of intermediate ages which at first sight seem to have much more in common with our own." He attempts to shew that "freedom is everywhere older than bondage," "toleration than intolerance." "No book could possibly be more useful to students of our Constitutional history, or a more pleasant means of conveying information about it to the public at large."—SATURDAY REVIEW.

Galileo.—THE PRIVATE LIFE OF GALILEO. Compiled principally from his Correspondence and that of his eldest daughter, Sister Maria Celeste, Nun in the Franciscan Convent of S. Matthew in Arcetri. With Portrait. Crown 8vo. 7s. 6d.

It has been the endeavour of the compiler to place before the reader a plain, ungarbled statement of facts; and, as a means to this end, to allow Galileo, his friends, and his judges to speak for themselves as far as possible. All the best authorities have been made use of, and all the materials which exist for a biography have been in this volume put into a symmetrical form. The result is a most touching picture skilfully arranged of the great heroic man of science and his devoted daughter, whose letters are full of the deepest reverential love and trust, amply repaid by the noble soul. The SATURDAY REVIEW *says of the book, "It is not so much the philosopher as the man who is seen in this simple and life-like sketch, and the hand which portrays the features and actions is mainly that of one who had studied the subject the closest and the most intimately. This little volume has done much within its slender compass to prove the depth and tenderness of Galileo's heart."*

Gladstone (Right Hon. W. E., M.P.)—JUVENTUS MUNDI. The Gods and Men of the Heroic Age. Crown 8vo. cloth. With Map. 10s. 6d. Second Edition.

This work of Mr. Gladstone deals especially with the historic element in Homer, expounding that element and furnishing by its aid a full account of the Homeric men and the Homeric religion. It starts, after the introductory chapter, with a discussion of the several races then existing in Hellas, including the influence of the Phœnicians and Egyptians. It contains chapters on the Olympian system, with its several deities; on the Ethics and the Polity of the Heroic age; on the Geography of Homer; on the characters of the Poems; presenting, in fine, a view of primitive life and primitive society as found in the poems of Homer. To this New Edition various additions have been made. "Seldom," says the ATHENÆUM, *"out of the great poems themselves, have these Divinities looked so majestic and respectable. To read these brilliant details is like standing on the Olympian threshold and gazing at the ineffable brightness within."*

"*There is,*" *according to the* WESTMINSTER REVIEW, "*probably no other writer now living who could have done the work of this book. . . . It would be difficult to point out a book that contains so much fulness of knowledge along with so much freshness of perception and clearness of presentation.*"

GOETHE AND MENDELSSOHN (1821—1831). From the German of Dr. KARL MENDELSSOHN, Son of the Composer, by M. E. VON GLEHN. From the Private Diaries and Home-Letters of Mendelssohn, with Poems and Letters of Goethe never before printed. Also with two New and Original Portraits, Facsimiles, and Appendix of Twenty Letters hitherto unpublished. Crown 8vo. 5s.

This little volume is full of interesting details about Mendelssohn from his twelfth year onwards, and especially of his intimate and frequent intercourse with Goethe. It is an episode of Wiemar's golden days which we see before us—old age and fame hand in hand with youth in its aspiring efforts; the aged poet fondling the curls of the little musician and calling to him in playful and endearing accents "to make a little noise for him, and awaken the winged spirits that have so long lain slumbering." Here will be found letters and reports of conversations between the two, touching on all subjects, human and divine—Music, Æsthetics, Art, Poetry, Science, Morals, and "the profound and ancient problem of human life," as well as reminiscences of celebrated men with whom the great composer came in contact. The letters appended give, among other matters, some interesting glimpses into the private life of Her Majesty Queen Victoria and the late Prince Albert. The two well-executed engravings show Mendelssohn as a beautiful boy of twelve years.

Guizot.—M. DE BARANTE, a Memoir, Biographical and Autobiographical. By M. GUIZOT. Translated by the Author of "JOHN HALIFAX, GENTLEMAN." Crown 8vo. 6s. 6d.

"*It is scarcely necessary to write a preface to this book. Its lifelike, portrait of a true and great man, painted unconsciously by himself in his letters and autobiography, and retouched and completed by the tender hand of his surviving friend—the friend of a lifetime—is sure, I think, to be appreciated in England as it was in France, where it appeared in the Revue de Deux Mondes. Also, I believe every thoughtful mind will enjoy its clear reflections of French and European politics and history for* the last seventy years, and the curious light thus thrown upon many present events and combinations of circumstances."—PREFACE. "The highest purposes of both history and biography are answered by a memoir so lifelike, so faithful, and so philosophical."—BRITISH QUARTERLY REVIEW. "This eloquent memoir, which for tenderness, gracefulness, and vigour, might be placed on the same shelf with Tacitus' Life of Agricola. . . . Mrs. Craik has rendered the language of Guizot in her own sweet translucent English."—DAILY NEWS.

Hole.—A GENEALOGICAL STEMMA OF THE KINGS OF ENGLAND AND FRANCE. By the Rev. C. HOLE M.A., Trinity College, Cambridge. On Sheet, 1s.

The different families are printed in distinguishing colours, thus facilitating reference.

Hozier (H. M.)—Works by CAPTAIN HENRY M. HOZIER, late Assistant Military Secretary to Lord Napier of Magdala.

THE SEVEN WEEKS' WAR; Its Antecedents and Incidents. New and Cheaper Edition. With New Preface, Maps, and Plans. Crown 8vo. 6s.

This account of the brief but momentous Austro-Prussian War of 1866 claims consideration as being the product of an eye-witness of some of its most interesting incidents. The author has attempted to ascertain and to advance facts. Two maps are given, one illustrating the operations of the Army of the Maine, and the other the operations from Königgrätz. In the Prefatory Chapter to this edition, events resulting from the war of 1866 are set forth, and the current of European history traced down to the recent Franco-Prussian war, a natural consequence of the war whose history is narrated in this volume. "Mr. Hozier added to the knowledge of military operations and of languages, which he had proved himself to possess, a ready and skilful pen, and excellent faculties of observation and description. . . . All that Mr. Hozier saw of the great events of the war—and he saw a large share of them—he describes in clear and vivid language."—SATURDAY REVIEW. "Mr. Hozier's volumes deserve to take a permanent place in the literature of the Seven Weeks' War."—PALL MALL GAZETTE.

Hozier (H. M.)—*continued.*

THE BRITISH EXPEDITION TO ABYSSINIA. Compiled from Authentic Documents. 8vo. 9s.

Several accounts of the British Expedition have been published. They have, however, been written by those who have not had access to those authentic documents, which cannot be collected directly after the termination of a campaign. The endeavour of the author of this sketch has been to present to readers a succinct and impartial account of an enterprise which has rarely been equalled in the annals of war. "This," says the SPECTATOR, *"will be the account of the Abyssinian Expedition for professional reference, if not for professional reading. Its literary merits are really very great."*

THE INVASIONS OF ENGLAND. A History of the Past, with Lessons for the Future. [In the press.

Huyshe (Captain G. L.)—THE RED RIVER EXPEDITION. By Captain G. L. HUYSHE, Rifle Brigade, late on the Staff of Colonel Sir GARNET WOLSELEY. With Maps. 8vo. 10s. 6d.

This account has been written in the hope of directing attention to the successful accomplishment of an expedition which was attended with more than ordinary difficulties. The author has had access to the official documents of the Expedition, and has also availed himself of the reports on the line of route published by Mr. Dawson, C.E., and by the Typographical Department of the War Office. The statements made may therefore be relied on as accurate and impartial. The endeavour has been made to avoid tiring the general reader with dry details of military movements, and yet not to sacrifice the character of the work as an account of a military expedition. The volume contains a portrait of President Louis Riel, and Maps of the route. The ATHENÆUM *calls it "an enduring authentic record of one of the most creditable achievements ever accomplished by the British Army."*

INSIDE PARIS DURING THE SIEGE. By an OXFORD GRADUATE. Crown 8vo. 7s. 6d.

This volume consists of the diary kept by a gentleman who lived in Paris during the whole of its siege by the Prussians. He had many facilities for coming in contact with men of all parties and of all classes, and ascertain-

ing the actual motives which animated them, and their real ultimate aims. These facilities he took advantage of, and in his diary, day by day, carefully recorded the results of his observations, as well as faithfully but graphically photographed the various incidents of the siege which came under his own notice, the actual condition of the besieged, the sayings and doings, the hopes and fears of the people among whom he freely moved. In the Appendix is an exhaustive and elaborate account of the Organization of the Republican party, sent to the author by M. Jules Andrieu; and a translation of the Manifesto of the Commune to the People of England, dated April 19, 1871. "The author tells his story admirably. The Oxford Graduate seems to have gone everywhere, heard what everyone had to say, and so been able to give us photographs of Paris life during the siege which we have not had from any other source."—SPECTATOR. "He has written brightly, lightly, and pleasantly, yet in perfect good taste."—SATURDAY REVIEW.

Irving.—THE ANNALS OF OUR TIME. A Diurnal of Events, Social and Political, Home and Foreign, from the Accession of Queen Victoria to the Peace of Versailles. By JOSEPH IRVING. Third Edition. 8vo. half-bound. 16s.

Every occurrence, metropolitan or provincial, home or foreign, which gave rise to public excitement or discussion, or became the starting point for new trains of thought affecting our social life, has been judged proper matter for this volume. In the proceedings of Parliament, an endeavour has been made to notice all those Debates which were either remarkable as affecting the fate of parties, or led to important changes in our relations with Foreign Powers. Brief notices have been given of the death of all noteworthy persons. Though the events are set down day by day in their order of occurrence, the book is, in its way, the history of an important and well-defined historic cycle. In these 'Annals,' the ordinary reader may make himself acquainted with the history of his own time in a way that has at least the merit of simplicity and readiness; the more cultivated student will doubtless be thankful for the opportunity given him of passing down the historic stream undisturbed by any other theoretical or party feeling than what he himself has at hand to explain the philosophy of our national story. A complete and useful Index is appended. The Table of Administrations is designed to assist the reader in following the various political changes noticed in their chronological order in the 'Annals.'— In the new edition all errors and omissions have been rectified, 300 pages been added, and as many as 46 occupied by an impartial exhibition of the

wonderful series of events marking the latter half of 1870. "*We have before us a trusty and ready guide to the events of the past thirty years, available equally for the statesman, the politician, the public writer, and the general reader. If Mr. Irving's object has been to bring before the reader all the most noteworthy occurrences which have happened since the beginning of her Majesty's reign, he may justly claim the credit of having done so most briefly, succinctly, and simply, and in such a manner, too, as to furnish him with the details necessary in each case to comprehend the event of which he is in search in an intelligent manner.*"
—TIMES.

Kingsley (Canon).—Works by the Rev. CHARLES KINGSLEY, M.A., Rector of Eversley and Canon of Chester. (For other Works by the same Author, *see* THEOLOGICAL and BELLES LETTRES Catalogues.)

ON THE ANCIEN RÉGIME as it existed on the Continent before the FRENCH REVOLUTION. Three Lectures delivered at the Royal Institution. Crown 8vo. 6s.

These three lectures discuss severally (1) *Caste,* (2) *Centralization,* (3) *The Explosive Forces by which the Revolution was superinduced. The Preface deals at some length with certain political questions of the present day.*

AT LAST: A CHRISTMAS in the WEST INDIES. With nearly Fifty Illustrations. New and Cheaper Edition. Crown 8vo.

Mr. Kingsley's dream of forty years was at last fulfilled, when he started on a Christmas expedition to the West Indies, for the purpose of becoming personally acquainted with the scenes which he has so vividly described in "*Westward Ho!*" *These two volumes are the journal of his voyage. Records of natural history, sketches of tropical landscape, chapters on education, views of society, all find their place in a work written, so to say, under the inspiration of Sir Walter Raleigh and the other adventurous men who three hundred years ago disputed against Philip II. the possession of the Spanish Main.* "*We can only say that Mr. Kingsley's account of a* '*Christmas in the West Indies*' *is in every way worthy to be classed among his happiest productions.*"—STANDARD.

THE ROMAN AND THE TEUTON. A Series of Lectures delivered before the University of Cambridge. 8vo. 12s.

CONTENTS :—*Inaugural Lecture ; The Forest Children ; The Dying Empire ; The Human Deluge ; The Gothic Civilizer; Dietrich's End; The Nemesis of the Goths ; Paulus Diaconus ; The Clergy and the Heathen ; The Monk a Civilizer ; The Lombard Laws ; The Popes and the Lombards ; The Strategy of Providence.* "He has rendered," says the NONCONFORMIST, "*good service and shed a new lustre on the chair of Modern History at Cambridge . . . . He has thrown a charm around the work by the marvellous fascinations of his own genius, brought out in strong relief those great principles of which all history is a revelation, lighted up many dark and almost unknown spots, and stimulated the desire to understand more thoroughly one of the greatest movements in the story of humanity.*"

Kingsley (Henry, F.R.G.S.)—For other Works by same Author, *see* BELLES LETTRES CATALOGUE.

TALES OF OLD TRAVEL. Re-narrated by HENRY KINGSLEY, F.R.G.S. With *Eight Illustrations* by HUARD. Third Edition. Crown 8vo. 6*s*.

In this volume Mr. Henry Kingsley re-narrates, at the same time preserving much of the quaintness of the original, some of the most fascinating tales of travel contained in the collections of Hakluyt and others. The CONTENTS *are—Marco Polo ; The Shipwreck of Pelsart ; The Wonderful Adventures of Andrew Battel ; The Wanderings of a Capuchin ; Peter Carder ; The Preservation of the "Terra Nova ;" Spitzbergen ; D'Ermenonville's Acclimatization Adventure ; The Old Slave Trade ; Miles Philips ; The Sufferings of Robert Everard ; John Fox ; Alvaro Nunez ; The Foundation of an Empire.* "*We know no better book for those who want knowledge or seek to refresh it. As for the 'sensational,' most novels are tame compared with these narratives.*"—ATHENÆUM. "*Exactly the book to interest and to do good to intelligent and high-spirited boys.*"—LITERARY CHURCHMAN.

Labouchere.—DIARY OF THE BESIEGED RESIDENT IN PARIS. Reprinted from the *Daily News*, with several New Letters and Preface. By HENRY LABOUCHERE. *Third Edition*. Crown 8vo. 6*s*.

" The ' *Diary of a Besieged Resident in Paris* ' *will certainly form one of the most remarkable records of a momentous episode in history.*"—SPECTATOR. "*There is an entire absence of affectation in this writer which vastly commends him to us."*—PALL MALL GAZETTE. *"On the whole, it does not seem likely that the 'besieged' will be superseded in his self-assumed function by any subsequent chronicler."*—BRITISH QUARTERLY REVIEW. *"Very smartly written."*—VANITY FAIR.

Macmillan (Rev. Hugh).—For other Works by same Author, see THEOLOGICAL and SCIENTIFIC CATALOGUES.

HOLIDAYS ON HIGH LANDS; or, Rambles and Incidents in search of Alpine Plants. Crown 8vo. cloth. 6s.

The aim of this book is to impart a general idea of the origin, character, and distribution of those rare and beautiful Alpine plants which occur on the British hills, and which are found almost everywhere on the lofty mountain chains of Europe, Asia, Africa, and America. The information the author has to give is conveyed in untechnical language, in a setting of personal adventure, and associated with descriptions of the natural scenery and the peculiarities of the human life in the midst of which the plants were found. By this method the subject is made interesting to a very large class of readers. "Botanical knowledge is blended with a love of nature, a pious enthusiasm, and a rich felicity of diction not to be met with in any works of kindred character, if we except those of Hugh Miller."—TELEGRAPH. *"Mr. M.'s glowing pictures of Scandinavian scenery."*—SATURDAY REVIEW.

Martin (Frederick).—THE STATESMAN'S YEAR-BOOK: See p. 41 of this Catalogue.

Martineau.—BIOGRAPHICAL SKETCHES, 1852—1868. By HARRIET MARTINEAU. Third and Cheaper Edition, with New Preface. Crown 8vo. 6s.

A Collection of Memoirs under these several sections:—(1) *Royal,* (2) *Politicians,* (3) *Professional,* (4) *Scientific,* (5) *Social,* (6) *Literary. These Memoirs appeared originally in the columns of the* DAILY NEWS. *"Miss Martineau's large literary powers and her fine intellectual training make these little sketches more instructive, and constitute them more genuinely works of art, than many more ambitious and diffuse biographies."*—FORTNIGHTLY REVIEW. *"Each memoir is a complete digest of a celebrated life, illuminated by the flood of searching light which streams from the gaze of an acute but liberal mind."*—MORNING STAR.

Masson (David).—For other Works by same Author, *see* PHILOSOPHICAL and BELLES LETTRES CATALOGUES.

LIFE OF JOHN MILTON. Narrated in connection with the Political, Ecclesiastical, and Literary History of his Time. By DAVID MASSON, M.A., LL.D., Professor of Rhetoric and English Literature in the University of Edinburgh. Vol. I. with Portraits. 8vo. 18*s.* Vol. II., 1638—1643. 8vo. 16*s.* Vol. III. in the press.

This work is not only a Biography, but also a continuous Political, Ecclesiastical, and Literary History of England through Milton's whole time. In order to understand Milton, his position, his motives, his thoughts by himself, his public words to his countrymen, and the probable effect of those words, it was necessary to refer largely to the History of his Time, not only as it is presented in well-known books, but as it had to be rediscovered by express and laborious investigation in original and forgotten records: thus of the Biography, a History grew: not a mere popular compilation, but a work of independent search and method from first to last, which has cost more labour by far than the Biography. The second volume is so arranged that the reader may select or omit either the History or Biography. The NORTH BRITISH REVIEW, *speaking of the first volume of this work said,* "*The Life of Milton is here written once for all.*" *The* NONCONFORMIST, *in noticing the second volume, says,* "*Its literary excellence entitles it to take its place in the first ranks of our literature, while the whole style of its execution marks it as the only book that has done anything like adequate justice to one of the great masters of our language, and one of our truest patriots, as well as our greatest epic poet.*"

Mayor (J. E. B.)—WORKS Edited By JOHN E. B. MAYOR, M.A., Kennedy Professor of Latin at Cambridge.

CAMBRIDGE IN THE SEVENTEENTH CENTURY. Part II. Autobiography of Matthew Robinson. Fcap. 8vo. 5*s.* 6*d.*

This is the second of the Memoirs illustrative of "*Cambridge in the Seventeenth Century*" *that of Nicholas Farrar having preceded it. It gives a lively picture of England during the Civil Wars, the most important crisis of our national life; it supplies materials for the history of the University and our Endowed Schools, and gives us a view of country clergy at a time when they are supposed to have been, with scarce an ex-*

ception, scurrilous sots. Mr. Mayor has added a collection of extracts and documents relating to the history of several other Cambridge men of note belonging to the same period, all, like Robinson, of Nonconformist leanings.

LIFE OF BISHOP BEDELL. By his SON. Fcap. 8vo. 3s. 6d.

This is the third of the Memoirs illustrative of " Cambridge in the 17th Century." The life of the Bishop of Kilmore here printed for the first time is preserved in the Tanner MSS., and is preliminary to a larger one to be issued shortly.

Mitford (A. B.)—TALES OF OLD JAPAN. By A. B. MITFORD, Second Secretary to the British Legation in Japan. With upwards of 30 Illustrations, drawn and cut on Wood by Japanese Artists. Two Vols. crown 8vo. 21s.

*Under the influence of more enlightened ideas and of a liberal system of policy, the old Japanese civilization is fast disappearing, and will, in a few years, be completely extinct. It was important, therefore, to preserve as far as possible trustworthy records of a state of society which, although venerable from its antiquity, has for Europeans the dawn of novelty; hence the series of narratives and legends translated by Mr. Mitford, and in which the Japanese are very judiciously left to tell their own tale. The two volumes comprise not only stories and episodes illustrative of Asiatic superstitions, but also three sermons. The preface, appendices, and notes explain a number of local peculiarities; the thirty-one woodcuts are the genuine work of a native artist, who, unconsciously of course, has adopted the process first introduced by the early German masters. " These very original volumes will always be interesting as memorials of a most exceptional society, while regarded simply as tales, they are sparkling, sensational, and dramatic, and the originality of their ideas and the quaintness of their language give them a most captivating piquancy. The illustrations are extremely interesting, and for the curious in such matters have a special and particular value."—*PALL MALL GAZETTE.

Morley (John).—EDMUND BURKE, a Historical Study. By JOHN MORLEY, B.A. Oxon. Crown 8vo. 7s. 6d.

" The style is terse and incisive, and brilliant with epigram and point. It contains pithy aphoristic sentences which Burke himself would not have disowned. Its sustained power of reasoning, its wide sweep of observation and reflection, its elevated ethical and social tone, stamp it as a work of

high excellence."—SATURDAY REVIEW. *"A model of compact condensation. We have seldom met with a book in which so much matter was compressed into so limited a space."*—PALL MALL GAZETTE. *"An essay of unusual effort."*—WESTMINSTER REVIEW.

Morison.—THE LIFE AND TIMES OF SAINT BERNARD, Abbot of Clairvaux. By JAMES COTTER MORISON, M.A. Cheaper Edition. Crown 8vo. 4s. 6d.

The PALL MALL GAZETTE *calls this "one of the best contributions in our literature towards a vivid, intelligent, and worthy knowledge of European interests and thoughts and feelings during the twelfth century. A delightful and instructive volume, and one of the best products of the modern historic spirit." "A work," says the* NONCONFORMIST, *"of great merit and value, dealing most thoroughly with one of the most interesting characters, and one of the most interesting periods, in the Church history of the Middle Ages. Mr. Morison is thoroughly master of his subject, and writes with great discrimination and fairness, and in a chaste and elegant style." The* SPECTATOR *says it is "not only distinguished by research and candour, it has also the great merit of never being dull."*

Napoleon I.—THE HISTORY OF NAPOLEON THE FIRST. By P. LANFREY. Translated with the sanction of the Author. Vols. I. and II. 8vo. 12s. each.

M. Lanfrey's History of Napoleon has taken its place in French literature as the standard history of the period with which it is concerned, occupying a place similar to that occupied by such histories as those of Palgrave and Froude in England. The author has written his history under the belief that the time has come to form a clear-sighted estimate of Napoleon's Life and Character, uninfluenced either by that profound hatred or profound attachment by which previous historians have allowed their judgment to be biased. The QUARTERLY REVIEW, *speaking of the French edition, says that its scope and tendency throughout are to disabuse the public mind of a cherished error, and at least compel a discriminating judgment from posterity. . . . . . A startling amount of new material for Napoleonic history has been brought to light within a few years in the shape of Memoirs, Letters, and Despatches, and the whole of these have been subjected to the minutest investigation by M. Lanfrey, who has thereby been enabled to light up his narrative with numerous traits and touches that give it an air of novelty, even when the scene is* crowded with familiar faces and the main action is well known." No one who wishes to understand clearly and thoroughly the History of Napoleon and his time can afford to omit reading the History of M. Lanfrey. "An excellent translation of a work on every ground deserving to be translated. It is unquestionably and immeasurably the best that has been produced. It is in fact the only work to which we can turn for an accurate and trustworthy narrative of that extraordinary career."—SATURDAY REVIEW.

Palgrave (Sir F.)—HISTORY OF NORMANDY AND OF ENGLAND. By Sir FRANCIS PALGRAVE, Deputy Keeper of Her Majesty's Public Records. Completing the History to the Death of William Rufus. Four Vols. 8vo. £4 4s.

Volume I. General Relations of Mediæval Europe—The Carlovingian Empire—The Danish Expeditions in the Gauls—And the Establishment of Rollo. Volume II. The Three First Dukes of Normandy; Rollo, Guillaume Longue-Épée, and Richard Sans-Peur—The Carlovingian line supplanted by the Capets. Volume III. Richard Sans-Peur— Richard Le-Bon—Richard III.—Robert Le Diable—William the Conqueror. Volume IV. William Rufus—Accession of Henry Beauclerc. It is needless to say anything to recommend this work of a lifetime to all students of history; it is, as the SPECTATOR *says, "perhaps the greatest single contribution yet made to the authentic annals of this country," and "must," says the* NONCONFORMIST, *"always rank among our standard authorities."*

Palgrave (W. G.)—A NARRATIVE OF A YEAR'S JOURNEY THROUGH CENTRAL AND EASTERN ARABIA, 1862-3. By WILLIAM GIFFORD PALGRAVE, late of the Eighth Regiment Bombay N. I. Sixth Edition. With Maps, Plans, and Portrait of Author, engraved on steel by Jeens. Crown 8vo. 6s.

"The work is a model of what its class should be; the style restrained, the narrative clear, telling us all we wish to know of the country and people visited, and enough of the author and his feelings to enable us to trust ourselves to his guidance in a tract hitherto untrodden, and dangerous in more senses than one. . . He has not only written one of the best books on the Arabs and one of the best books on Arabia, but he has done so in a manner that must command the respect no less than the admiration of his fellow-countrymen."—FORTNIGHTLY REVIEW. " Considering the extent *of our previous ignorance, the amount of his achievements, and the importance of his contributions to our knowledge, we cannot say less of him than was once said of a far greater discoverer—Mr. Palgrave has indeed given a new world to Europe."*—PALL MALL GAZETTE.

Prichard.—THE ADMINISTRATION OF INDIA. From 1859 to 1868. The First Ten Years of Administration under the Crown. By ILTUDUS THOMAS PRICHARD, Barrister-at-Law. Two Vols. Demy 8vo. With Map. 21s.

In these volumes the author has aimed to supply a full, impartial, and independent account of British India between 1859 and 1868—which is in many respects the most important epoch in the history of that country that the present century has seen. "It has the great merit that it is not exclusively devoted, as are too many histories, to military and political details, but enters thoroughly into the more important questions of social history. We find in these volumes a well-arranged and compendious reference to almost all that has been done in India during the last ten years; and the most important official documents and historical pieces are well selected and duly set forth."—SCOTSMAN. *"It is a work which every Englishman in India ought to add to his library."*—STAR OF INDIA.

Robinson (H. Crabb)—THE DIARY, REMINISCENCES, AND CORRESPONDENCE, OF HENRY CRABB ROBINSON, Barrister-at-Law. Selected and Edited by THOMAS SADLER, Ph.D. With Portrait. Third and Cheaper Edition. Two Vols. Crown 8vo. 16s.

The DAILY NEWS *says: "The two books which are most likely to survive change of literary taste, and to charm while instructing generation after generation, are the 'Diary' of Pepys and Boswell's 'Life of Johnson.' The day will come when to these many will add the 'Diary of Henry Crabb Robinson.' Excellences like those which render the personal revelations of Pepys and the observations of Boswell such pleasant reading abound in this work .... In it is to be found something to suit every taste and inform every mind. For the general reader it contains much light and amusing matter. To the lover of literature it conveys information which he will prize highly on account of its accuracy and rarity. The student of social life will gather from it many valuable hints whereon to base theories as to the effects on English society of the progress of civilization. For these and other reasons this 'Diary' is a work to which a hearty welcome should be accorded."*

Rogers (James E. Thorold).—HISTORICAL GLEANINGS : A Series of Sketches. Montague, Walpole, Adam Smith, Cobbett. By Prof. ROGERS. Crown 8vo. 4*s.* 6*d.* Second Series. Wiklif, Laud, Wilkes, and Horne Tooke. Crown 8vo. 6*s.*

Professor Rogers's object in these sketches, which are in the form of Lectures, is to present a set of historical facts, grouped round a principal figure. The author has aimed to state the social facts of the time in which the individual whose history is handled took part in public business. It is from sketches like these of the great men who took a prominent and influential part in the affairs of their time that a clear conception of the social and economical condition of our ancestors can be obtained. History learned in this way is both instructive and agreeable. " *His Essays,*" *the* PALL MALL GAZETTE *says,* "*are full of interest, pregnant, thoughtful, and readable.*" " *They rank far above the average of similar performances,*" *says the* WESTMINSTER REVIEW.

Raphael.—RAPHAEL OF URBINO AND HIS FATHER GIOVANNI SANTI. By J. D. PASSAVANT, formerly Director of the Museum at Frankfort. With Twenty Permanent Photographs. Royal 8vo. Handsomely bound. 31*s.* 6*d.*

To the enlarged French edition of Passavant's Life of Raphael, that painter's admirers have turned whenever they have sought information, and it will doubtless remain for many years the best book of reference on all questions pertaining to the great painter. The present work consists of a translation of those parts of Passavant's volumes which are most likely to interest the general reader. Besides a complete life of Raphael, it contains the valuable descriptions of all his known paintings, and the Chronological Index, which is of so much service to amateurs who wish to study the progressive character of his works. The Illustrations by Woodbury's new permanent process of photography, are taken from the finest engravings that could be procured, and have been chosen with the intention of giving examples of Raphael's various styles of painting. The SATURDAY REVIEW *says of them,* " *We have seen not a few elegant specimens of Mr. Woodbury's new process, but we have seen none that equal these.*"

Somers (Robert).—THE SOUTHERN STATES SINCE THE WAR. By ROBERT SOMERS. With Map. 8vo. 9*s.*

This work is the result of inquiries made by the author of all authorities competent to afford him information, and of his own observation during a lengthened sojourn in the Southern States, to which writers on America so seldom direct their steps. The author's object is to give some account of the condition of the Southern States under the new social and political system introduced by the civil war. He has here collected such notes of the progress of their cotton plantations, of the state of their labouring population and of their industrial enterprises, as may help the reader to a safe opinion of their means and prospects of development. He also gives such information of their natural resources, railways, and other public works, as may tend to show to what extent they are fitted to become a profitable field of enlarged immigration, settlement, and foreign trade. The volume contains many valuable and reliable details as to the condition of the Negro population, the state of Education and Religion, of Cotton, Sugar, and Tobacco Cultivation, of Agriculture generally, of Coal and Iron Mining, Manufactures, Trade, Means of Locomotion, and the condition of Towns and of Society. A large map of the Southern States by Messrs. W. and A. K. Johnston is appended, which shows with great clearness the Cotton, Coal, and Iron districts, the railways completed and projected, the State boundaries, and other important details. "Full of interesting and valuable information."—SATURDAY REVIEW.

Smith (Professor Goldwin).—THREE ENGLISH STATESMEN. See p. 41 of this Catalogue.

Tacitus.—THE HISTORY OF TACITUS, translated into English. By A. J. CHURCH, M.A. and W. J. BRODRIBB, M.A. With a Map and Notes. New Edition in the press.

The translators have endeavoured to adhere as closely to the original as was thought consistent with a proper observance of English idiom. At the same time it has been their aim to reproduce the precise expressions of the author. This work is characterised by the SPECTATOR *as " a scholarly and faithful translation."*

THE AGRICOLA AND GERMANIA. Translated into English by A. J. CHURCH, M.A. and W. J. BRODRIBB, M.A. With Maps and Notes. Extra fcap. 8vo. 2s. 6d.

The translators have sought to produce such a version as may satisfy scholars who demand a faithful rendering of the original, and English readers who are offended by the baldness and frigidity which commonly disfigure translations. The treatises are accompanied by Introductions, Notes, Maps, and a chronological Summary. The ATHENÆUM *says of* this work that it is "*a version at once readable and exact, which may be perused with pleasure by all, and consulted with advantage by the classical student;*" and the PALL MALL GAZETTE says, "*What the editors have attempted to do, it is not, we think probable, that any living scholars could have done better.*"

Taylor (Rev. Isaac).—WORDS AND PLACES. See p. 49 of this Catalogue.

Trench (Archbishop).—For other Works by the same Author, see THEOLOGICAL and BELLES LETTRES CATALOGUES, and p. 50 of this Catalogue.

GUSTAVUS ADOLPHUS IN GERMANY, and other Lectures on the Thirty Years' War. By R. CHENEVIX TRENCH, D.D., Archbishop of Dublin. Second Edition, revised and enlarged. Fcap. 8vo. 4s.

The lectures contained in this volume form rather a new book than a new edition, for on the two lectures published by the Author several years ago, so many changes and additions have been made, as to make the work virtually a new one. Besides three lectures of the career of Gustavus in Germany and during the Thirty Years' War, there are other two, one on "Germany during the Thirty Years' War," and another on Germany after that War. The work will be found not only interesting and instructive in itself, but will be found to have some bearing on events connected with the recent European War.

Trench (Mrs. R.)—Remains of the late MRS. RICHARD TRENCH. Being Selections from her Journals, Letters, and other Papers. Edited by ARCHBISHOP TRENCH. New and Cheaper Issue, with Portrait. 8vo. 6s.

Contains Notices and Anecdotes illustrating the social life of the period.—extending over a quarter of a century (1799—1827). It includes also Poems and other miscellaneous pieces by Mrs. Trench.

Wallace.—Works by ALFRED RUSSEL WALLACE. For other Works by same Author, see SCIENTIFIC CATALOGUE.

Dr. Hooker, in his address to the British Association, spoke thus of the author:—"Of Mr. Wallace and his many contributions to philosophical biology it is not easy to speak without enthusiasm; for, putting aside their great merits, he, throughout his writings, with a modesty as rare as I

Wallace (A. R.)—*continued.*

believe it to be unconscious, forgets his own unquestioned claim to the honour of having originated, independently of Mr. Darwin, the theories which he so ably defends."

A NARRATIVE OF TRAVELS ON THE AMAZON AND RIO NEGRO, with an Account of the Native Tribes, and Observations on the Climate, Geology, and Natural History of the Amazon Valley. With a Map and Illustrations. 8vo. 12*s.*

Mr. Wallace is acknowledged as one of the first of modern travellers and naturalists. This, his earliest work, will be found to possess many charms for the general reader, and to be full of interest to the student of natural history.

THE MALAY ARCHIPELAGO: the Land of the Orang Utan and the Bird of Paradise. A Narrative of Travel with Studies of Man and Nature. With Maps and Illustrations. Third and Cheaper Edition. Crown 8vo. 7*s.* 6*d.*

" The result is a vivid picture of tropical life, which may be read with unflagging interest, and a sufficient account of his scientific conclusions to stimulate our appetite without wearying us by detail. In short, we may safely say that we have never read a more agreeable book of its kind."—SATURDAY REVIEW. *" His descriptions of scenery, of the people and their manners and customs, enlivened by occasional amusing anecdotes, constitute the most interesting reading we have taken up for some time."*—STANDARD.

Ward (Professor).—THE HOUSE OF AUSTRIA IN THE THIRTY YEARS' WAR. Two Lectures, with Notes and Illustrations. By ADOLPHUS W. WARD, M.A., Professor of History in Owens College, Manchester. Extra fcap. 8vo. 2*s.* 6*d.*

These two Lectures were delivered in February, 1869, *at the Philosophical Institution, Edinburgh, and are now published with Notes and Illustrations. " We have never read,"* says the SATURDAY REVIEW, *" any lectures which bear more thoroughly the impress of one who has a true and vigorous grasp of the subject in hand." " They are,"* the SCOTSMAN says, *"the fruit of much labour and learning, and it would be difficult to compress into a hundred pages more information."*

Ward (J.).—EXPERIENCES OF A DIPLOMATIST. Being recollections of Germany founded on Diaries kept during the years 1840—1870. By JOHN WARD, C.B., late H.M. Minister-Resident to the Hanse Towns. 8vo. 10s. 6d.

Mr. Ward's recollections extend back even to 1830. From his official position as well as from other circumstances he had many opportunities of coming in contact with eminent men of all ranks and all professions on the Continent. His book, while it contains much that throws light on the history of the long and important period with which it is concerned, is full of reminiscences of such men as Arrivabene, King Leopold, Frederick William IV., his Court and Ministers, Humboldt, Bunsen, Raumer, Ranke, Grimm, Palmerston, Sir de Lacy Evans, Cobden, Mendelssohn, Cardinal Wiseman, Prince Albert, the Prince and Princess of Wales, Lord Russell, Bismarck, Mdlle. Tietjens, and many other eminent Englishmen and foreigners.

Warren.—AN ESSAY ON GREEK FEDERAL COINAGE. By the Hon. J. LEICESTER WARREN, M.A. 8vo. 2s. 6d.

The present essay is an attempt to illustrate Mr. Freeman's Federal Government by evidence deduced from the coinage of the times and countries therein treated of.

Wedgwood.—JOHN WESLEY AND THE EVANGELICAL REACTION of the Eighteenth Century. By JULIA WEDGWOOD. Crown 8vo. 8s. 6d.

This book is an attempt to delineate the influence of a particular man upon his age. The background to the central figure is treated with considerable minuteness, the object of representation being not the vicissitude of a particular life, but that element in the life which impressed itself on the life of a nation,—an element which cannot be understood without a study of aspects of national thought which on a superficial view might appear wholly unconnected with it. "*In style and intellectual power, in breadth of view and clearness of insight, Miss Wedgwood's book far surpasses all rivals.*"—ATHENÆUM. "*As a short account of the most remarkable movement in the eighteenth century, it must fairly be described as excellent.*"—PALL MALL GAZETTE.

Wilson.—A MEMOIR OF GEORGE WILSON, M.D., F.R.S.E., Regius Professor of Technology in the University of Edinburgh. By his SISTER. New Edition. Crown 8vo. 6s.

"*An exquisite and touching portrait of a rare and beautiful spirit.*"—GUARDIAN. "*He more than most men of whom we have lately read deserved a minute and careful biography, and by such alone could he be understood, and become loveable and influential to his fellow-men. Such a biography his sister has written, in which letters reach almost to the extent of a complete autobiography, with all the additional charm of being unconsciously such. We revere and admire the heart, and earnestly praise the patient tender hand, by which such a worthy record of the earth-story of one of God's true angel-men has been constructed for our delight and profit.*"—NONCONFORMIST.

Wilson (Daniel, LL.D.)—Works by DANIEL WILSON, LL.D., Professor of History and English Literature in University College, Toronto :—

PREHISTORIC ANNALS OF SCOTLAND. New Edition, with numerous Illustrations. Two Vols. demy 8vo. 36s.

One object aimed at when the book first appeared was to rescue archæological research from that limited range to which a too exclusive devotion to classical studies had given rise, and, especially in relation to Scotland, to prove how greatly more comprehensive and important are its native antiquities than all the traces of intruded art. The aim has been to a large extent effectually accomplished, and such an impulse given to archæological research, that in this new edition the whole of the work has had to be remodelled. Fully a third of it has been entirely re-written; and the remaining portions have undergone so minute a revision as to render it in many respects a new work. The number of pictorial illustrations has been greatly increased, and several of the former plates and woodcuts have been re-engraved from new drawings. This is divided into four Parts. Part I. deals with The Primeval or Stone Period : *Aboriginal Traces, Sepulchral Memorials, Dwellings, and Catacombs, Temples, Weapons, etc. etc.; Part II.* The Bronze Period : *The Metallurgic Transition, Primitive Bronze, Personal Ornaments, Religion, Arts, and Domestic Habits, with other topics; Part III.* The Iron Period : *The Introduction of Iron, The Roman Invasion, Strongholds, etc. etc.; Part IV.* The Christian Period : *Historical Data, the Norrie's Law Relics, Primitive and Mediæval Ecclesiology, Ecclesiastical and Miscellaneous Antiquities. The work is furnished with an elaborate Index.* "*One of the most interesting, learned, and elegant works we have seen for a long time.*"—WESTMINSTER REVIEW. "*The interest connected with this beautiful volume is not*

Wilson (Daniel, LL.D.)—*continued.*

limited to that part of the kingdom to which it is chiefly devoted; it will be consulted with advantage and gratification by all who have a regard for National Antiquities and for the advancement of scientific Archæology."— ARCHÆOLOGICAL JOURNAL.

PREHISTORIC MAN. New Edition, revised and partly re-written, with numerous Illustrations. One vol. 8vo. 21*s.*

This work, which carries out the principle of the preceding one, but with a wider scope, aims to "view Man, as far as possible, unaffected by those modifying influences which accompany the development of nations and the maturity of a true historic period, in order thereby to ascertain the sources from whence such development and maturity proceed. These researches into the origin of civilization have accordingly been pursued under the belief which influenced the author in previous inquiries that the investigations of the archæologist, when carried on in an enlightened spirit, are replete with interest in relation to some of the most important problems of modern science. To reject the aid of archæology in the progress of science, and especially of ethnological science, is to extinguish the lamp of the student when most dependent on its borrowed rays." A prolonged residence on some of the newest sites of the New World has afforded the author many opportunities of investigating the antiquities of the American Aborigines, and of bringing to light many facts of high importance in reference to primeval man. The changes in the new edition, necessitated by the great advance in Archæology since the first, include both reconstruction and condensation, along with considerable additions alike in illustration and in argument. "*We find,*" *says the* ATHENÆUM, "*the main idea of his treatise to be a pre-eminently scientific one,—namely, by archæological records to obtain a definite conception of the origin and nature of man's earliest efforts at civilization in the New World, and to endeavour to discover, as if by analogy, the necessary conditions, phases, and epochs through which man in the prehistoric stage in the Old World also must necessarily have passed." The* NORTH BRITISH REVIEW *calls it* "*a mature and mellow work of an able man; free alike from crotchets and from dogmatism, and exhibiting on every page the caution and moderation of a well-balanced judgment.*"

CHATTERTON: A Biographical Study. By DANIEL WILSON, LL.D., Professor of History and English Literature in University College, Toronto. Crown 8vo. 6*s.* 6*d.*

The author here regards Chatterton as a poet, not as a "mere resetter and defacer of stolen literary treasures." Reviewed in this light, he has found much in the old materials capable of being turned to new account: and to these materials research in various directions has enabled him to make some additions. He believes that the boy-poet has been misjudged, and that the biographies hitherto written of him are not only imperfect but untrue. While dealing tenderly, the author has sought to deal truthfully with the failings as well as the virtues of the boy: bearing always in remembrance, what has been too frequently lost sight of, that he was but a boy;—a boy, and yet a poet of rare power. The EXAMINER *thinks this "the most complete and the purest biography of the poet which has yet appeared." The* LITERARY CHURCHMAN *calls it "a most charming literary biography."*

Yonge (Charlotte M.)—Works by CHARLOTTE M. YONGE, Author of "The Heir of Redclyffe," &c. &c. :—

A PARALLEL HISTORY OF FRANCE AND ENGLAND: consisting of Outlines and Dates. Oblong 4to. 3s. 6d.

This tabular history has been drawn up to supply a want felt by many teachers of some means of making their pupils realize what events in the two countries were contemporary. A skeleton narrative has been constructed of the chief transactions in either country, placing a column between for what affected both alike, by which means it is hoped that young people may be assisted in grasping the mutual relation of events.

CAMEOS FROM ENGLISH HISTORY. From Rollo to Edward II. Extra fcap. 8vo. Second Edition, enlarged. 5s.

A SECOND SERIES, THE WARS IN FRANCE. Extra fcap. 8vo. 5s.

The endeavour has not been to chronicle facts, but to put together a series of pictures of persons and events, so as to arrest the attention, and give some individuality and distinctness to the recollection, by gathering together details of the most memorable moments. The "Cameos" are intended as a book for young people just beyond the elementary histories of England, and able to enter in some degree into the real spirit of events, and to be struck with characters and scenes presented in some relief. "Instead of dry details," says the NONCONFORMIST, *"we have living pictures, faithful, vivid, and striking."*

Young (Julian Charles, M.A.)—A MEMOIR OF CHARLES MAYNE YOUNG, Tragedian, with Extracts from his Son's Journal. By JULIAN CHARLES YOUNG, M.A. Rector of Ilmington. With Portraits and Sketches. *New and Cheaper Edition.* Crown 8vo. 7s. 6d.

Round this memoir of one who held no mean place in public estimation as a tragedian, and who, as a man, by the unobtrusive simplicity and moral purity of his private life, won golden opinions from all sorts of men, are clustered extracts from the author's Journals, containing many curious and interesting reminiscences of his father's and his own eminent and famous contemporaries and acquaintances, somewhat after the manner of H. Crabb Robinson's Diary. Every page will be found full both of entertainment and instruction. It contains four portraits of the tragedian, and a few other curious sketches. "In this budget of anecdotes, fables, and gossip, old and new, relative to Scott, Moore, Chalmers, Coleridge, Wordsworth, Croker, Mathews, the third and fourth Georges, Bowles, Beckford, Lockhart, Wellington, Peel, Louis Napoleon, D'Orsay, Dickens, Thackeray, Louis Blanc, Gibson, Constable, and Stanfield, etc. etc. the reader must be hard indeed to please who cannot find entertainment."— PALL MALL GAZETTE.

POLITICS, POLITICAL AND SOCIAL ECONOMY, LAW, AND KINDRED SUBJECTS.

Baxter.—NATIONAL INCOME: The United Kingdom. By R. DUDLEY BAXTER, M.A. 8vo. 3s. 6d.

The present work endeavours to answer systematically such questions as the following:—What are the means and aggregate wages of our labouring population; what are the numbers and aggregate profits of the middle classes; what the revenues of our great proprietors and capitalists; and what the pecuniary strength of the nation to bear the burdens annually falling upon us? What capital in land and goods and money is stored up for our subsistence, and for carrying out our enterprises? The author has collected his facts from every quarter and tested them in various ways, in order to make his statements and deductions valuable and trustworthy. Part I. of the work deals with the Classification of the Population into—*Chap. I.* The Income Classes; *Chap. II.* The Upper and Middle and Manual Labour Classes. *Part II. treats of the* Income of the United Kingdom, *divided into*—*Chap. III.* Upper and Middle Incomes; *Chap. IV.* Wages of the Manual Labour Classes—England and Wales; *Chap. V.* Income of Scotland; *Chap. VI.* Income of Ireland; *Chap. VII.* Income of the United Kingdom. *In the Appendix will be found many valuable and carefully compiled tables, illustrating in detail the subjects discussed in the text.*

Bernard.—FOUR LECTURES ON SUBJECTS CONNECTED WITH DIPLOMACY. By MONTAGUE BERNARD, M.A., Chichele Professor of International Law and Diplomacy, Oxford. 8vo. 9s.

These four Lectures deal with—I. "The Congress of Westphalia;" II. "Systems of Policy;" III. "Diplomacy, Past and Present;" IV. "The Obligations of Treaties."—"Singularly interesting lectures, so able, clear, and attractive."—SPECTATOR. "The author of these lectures is full of the knowledge which belongs to his subject, and has that power of clear and vigorous expression which results from clear and vigorous thought."—SCOTSMAN.

Bright (John, M.P.)—SPEECHES ON QUESTIONS OF PUBLIC POLICY. By the Right Hon. JOHN BRIGHT, M.P. Edited by Professor THOROLD ROGERS. Author's Popular Edition. Globe 8vo. 3s. 6d.

The speeches which have been selected for publication in these volumes possess a value, as examples of the art of public speaking, which no person will be likely to underrate. The speeches have been selected with a view of supplying the public with the evidence on which Mr. Bright's friends assert his right to a place in the front rank of English statesmen. They are divided into groups, according to their subjects. The editor has naturally given prominence to those subjects with which Mr. Bright has been specially identified, as, for example, India, America, Ireland, and Parliamentary Reform. But nearly every topic of great public interest on which Mr. Bright has spoken is represented in these volumes. "Mr. Bright's speeches will always deserve to be studied, as an apprenticeship to popular and parliamentary oratory; they will form materials for the history of our time, and many brilliant passages, perhaps some entire speeches, will really become a part of the living literature of England."—DAILY NEWS.

LIBRARY EDITION. Two Vols. 8vo. With Portrait. 25s.

Christie.—THE BALLOT AND CORRUPTION AND EXPENDITURE AT ELECTIONS, a Collection of Essays and Addresses of different dates. By W. D. CHRISTIE, C.B., formerly Her Majesty's Minister to the Argentine Confederation and to Brazil; Author of "Life of the First Earl of Shaftesbury." Crown 8vo. 4s. 6d.

Mr. Christie has been well known for upwards of thirty years as a strenuous and able advocate for the Ballot, both in his place in Parliament and elsewhere. The papers and speeches here collected *are six in number, exclusive of the Preface and Dedication to Professor Maurice, which contains many interesting historical details concerning the Ballot.* " *You have thought to greater purpose on the means of preventing electoral corruption, and are likely to be of more service in passing measures for that highly important end, than any other person that I could name.*"—J. S. Mill, in a published letter to the Author, May 1868.

Clarke.—EARLY ROMAN LAW. THE REGAL PERIOD. By E. C. CLARKE, M.A., of Lincoln's Inn, Barrister-at-Law, Lecturer in Law and late Fellow of Trinity College, Cambridge. Crown 8vo. 5s.

The beginnings of Roman Law are only noticed incidentally by Gaius or his paraphrasers under Justinian. They are, however, so important, that this attempt to set forth what is known or may be inferred about them, it is expected, will be found of much value. The method adopted by the author has been to furnish in the text of each section a continuous account of the subject in hand, ample quotations and references being appended in the form of notes. Most of the passages cited have been arrived at by independent reading of the original authority, the few others having been carefully verified. " *Mr. Clark has brought together a great mass of valuable matter in an accessible form.*"—SATURDAY REVIEW.

Corfield (Professor W. H.)—A DIGEST OF FACTS RELATING TO THE TREATMENT AND UTILIZATION OF SEWAGE. By W. H. CORFIELD, M.A., B.A., Professor of Hygiene and Public Health at University College, London. 8vo. 10s. 6d. Second Edition, corrected and enlarged.

In this edition the author has revised and corrected the entire work, and made many important additions. The headings of the eleven chapters are as follow:—I. "*Early Systems: Midden-Heaps and Cesspools.*" *II.* "*Filth and Disease—Cause and Effect.*" *III.* "*Improved Midden-Pits and Cesspools; Midden-Closets, Pail-Closets, etc.*" *IV.* "*The Dry-Closet Systems.*" *V.* "*Water-Closets.*" *VI.* "*Sewerage.*" *VII.* "*Sanitary Aspects of the Water-Carrying System.*" *VIII.* "*Value of Sewage; Injury to Rivers.*" *IX. Town Sewage; Attempts at Utilization.*" *X.* "*Filtration and Irrigation.*" *XI.* "*Influence of Sewage Farming on the Public*

Health." An abridged account of the more recently published researches on the subject will be found in the Appendices, while the Summary contains a concise statement of the views which the author himself has been led to adopt; references have been inserted throughout to show from what sources the numerous quotations have been derived, and an Index has been added. "Mr. Corfield's work is entitled to rank as a standard authority, no less than a convenient handbook, in all matters relating to sewage."—ATHENÆUM.

Fawcett.—Works by HENRY FAWCETT, M.A., M.P., Fellow of Trinity Hall, and Professor of Political Economy in the University of Cambridge :—

THE ECONOMIC POSITION OF THE BRITISH LABOURER. Extra fcp. 8vo. 5s.

This work formed a portion of a course of Lectures delivered by the author in the University of Cambridge, and he has deemed it advisable to retain many of the expositions of the elementary principles of Economic Science. In the Introductory Chapter the author points out the scope of the work and shows the vast importance of the subject in relation to the commercial prosperity and even the national existence of Britain. Then follow five chapters on "The Land Tenure of England," "Co-operation," "The Causes which regulate Wages," "Trade Unions and Strikes," and "Emigration." The EXAMINER *calls the work "a very scholarly exposition on some of the most essential questions of Political Economy;" and the* NONCONFORMIST *says "it is written with charming freshness, ease, and lucidity."*

MANUAL OF POLITICAL ECONOMY. Third and Cheaper Edition, with Two New Chapters. Crown 8vo. 10s. 6d.

In this treatise no important branch of the subject has been omitted, and the author believes that the principles which are therein explained will enable the reader to obtain a tolerably complete view of the whole science. Mr. Fawcett has endeavoured to show how intimately Political Economy is connected with the practical questions of life. For the convenience of the ordinary reader, and especially for those who may use the book to prepare themselves for examinations, he has prefixed a very detailed summary of Contents,

Fawcett (H.)—*continued.*

which may be regarded as an analysis of the work. The new edition has been so carefully revised that there is scarcely a page in which some improvement has not been introduced. The DAILY NEWS says: "*It forms one of the best introductions to the principles of the science, and to its practical applications in the problems of modern, and especially of English, government and society.*" "*The book is written throughout,*" says the EXAMINER, "*with admirable force, clearness, and brevity, every important part of the subject being duly considered.*"

PAUPERISM : ITS CAUSES AND REMEDIES. Crown 8vo. 5s. 6d.

In its number for March 11*th,* 1871, *the* SPECTATOR *said:* "*We wish Professor Fawcett would devote a little more of his time and energy to the practical consideration of that monster problem of Pauperism, for the treatment of which his economic knowledge and popular sympathies so eminently fit him.*" *The volume now published may be regarded as an answer to the above challenge. The seven chapters it comprises discuss the following subjects:*—*I.* "*Pauperism and the old Poor Law.*" *II.* "*The present Poor Law System.*" *III.* "*The Increase of Population.*" *IV.* "*National Education; its Economic and Social Effects.*" *V.* "*Co-partnership and Co-operation.*" *VI.* "*The English System of Land Tenure.*" *VII.* "*The Inclosure of Commons.*" *The* ATHENÆUM *calls the work* "*a repertory of interesting and well-digested information.*"

ESSAYS ON POLITICAL AND SOCIAL SUBJECTS. By PROFESSOR FAWCETT, M.P., and MILLICENT GARRETT FAWCETT. 8vo. 10s. 6d.

This volume contains fourteen papers, some of which have appeared in various journals and periodicals ; others have not before been published. They are all on subjects of great importance and universal interest, and the names of the two authors are a sufficient guarantee that each topic is discussed with full knowledge, great ability, clearness, and earnestness. The following are some of the titles :—"*Modern Socialism ;*" "*Free Education in its Economic Aspects ;*" "*Pauperism, Charity, and the Poor Law ;*" "*National Debt and National Prosperity ;*" "*What can be done for the*

Agricultural Labourers ;" " The Education of Women ;" " The Electoral Disabilities of Women ;" " The House of Lords." Each article is signed with the initials of its author. " In every respect a work of note and value. . . They will all repay the perusal of the thinking reader."—DAILY NEWS.

Fawcett (Mrs.)—POLITICAL ECONOMY FOR BEGINNERS. WITH QUESTIONS. By MILLICENT GARRETT FAWCETT. New Edition. 18mo. 2s. 6d.

In this little work are explained as briefly as possible the most important principles of Political Economy, in the hope that it will be useful to beginners, and perhaps be an assistance to those who are desirous of introducing the study of Political Economy to schools. In order to adapt the book especially for school use, questions have been added at the end of each chapter. In the new edition each page has been carefully revised, and at the end of each chapter, after the questions, a few little puzzles have been added, which will give interest to the book, and teach the learner to think for himself. The DAILY NEWS calls it "clear, compact, and comprehensive;" and the SPECTATOR says, "Mrs. Fawcett's treatise is perfectly suited to its purpose."

Freeman (E. A., M.A., D.C.L.)—HISTORY OF FEDERAL GOVERNMENT. See p. 7 of preceding HISTORICAL CATALOGUE.

Godkin (James).—THE LAND WAR IN IRELAND. A History for the Times. By JAMES GODKIN, Author of "Ireland and her Churches," late Irish Correspondent of the *Times*. 8vo. 12s.

A History of the Irish Land Question. " There is probably no other account so compendious and so complete."—FORTNIGHTLY REVIEW.

Guide to the Unprotected, in Every Day Matters Relating to Property and Income. By a BANKER'S DAUGHTER. Third Edition. Extra fcap. 8vo. 3s. 6d.

Many widows and single ladies, and all young people, on first possessing money of their own, are in want of advice when they *have commonplace business matters to transact. The author of this work writes for those who know* nothing. *Her aim throughout is to avoid all technicalities; to give plain and practical directions, not only as to what ought to be done, but how to do it.* "*Many an unprotected female will bless the head which planned and the hand which compiled this admirable little manual. . . . This book was very much wanted, and it could not have been better done.*"— MORNING STAR.

Hill.—CHILDREN OF THE STATE. THE TRAINING OF JUVENILE PAUPERS. By FLORENCE HILL. Extra fcap. 8vo. cloth. 5*s*.

In this work the author discusses the various systems adopted in this and other countries in the treatment of pauper children. The BIRMINGHAM DAILY GAZETTE *calls it* "*a valuable contribution to the great and important social question which it so ably and thoroughly discusses; and it must materially aid in producing a wise method of dealing with the Children of the State.*"

Historicus.—LETTERS ON SOME QUESTIONS OF INTERNATIONAL LAW. Reprinted from the *Times*, with considerable Additions. 8vo. 7*s*. 6*d*. Also, ADDITIONAL LETTERS. 8vo. 2*s*. 6*d*.

The author's intention in these Letters was to illustrate in a popular form clearly-established principles of law, or to refute, as occasion required, errors which had obtained a mischievous currency. He has endeavoured to establish, by sufficient authority, propositions which have been inconsiderately impugned, and to point out the various methods of reasoning which have led some modern writers to erroneous conclusions. The volume contains: Letters on "*Recognition;*" "*On the Perils of Intervention;*" "*The Rights and Duties of Neutral Nations;*" "*On the Law of Blockade;*" "*On Neutral Trade in Contraband of War;*" "*On Belligerent Violation of Neutral Rights;*" "*The Foreign Enlistment Act;*" "*The Right of Search;*" *extracts from letters on the Affair of the Trent; and a paper on the* "*Territoriality of the Merchant Vessel.*"—"*It is seldom that the doctrines of International Law on debateable points have been stated with more vigour, precision, and certainty.*"—SATURDAY REVIEW.

Jevons.—Works by W. STANLEY JEVONS, M.A., Professor of Logic and Political Economy in Owens College, Manchester. (For other Works by the same Author, *see* EDUCATIONAL and PHILOSOPHICAL CATALOGUES.)

THE COAL QUESTION: An Inquiry Concerning the Progress of the Nation, and the Probable Exhaustion of our Coal Mines. Second Edition, revised. 8vo. 10s. 6d.

> "*Day by day,*" *the author says,* "*it becomes more evident that the coal we happily possess in excellent quality and abundance is the mainspring of modern material civilization.*" *Geologists and other competent authorities have of late been hinting that the supply of coal is by no means inexhaustible, and as it is of vast importance to the country and the world generally to know the real state of the case, Professor Jevons in this work has endeavoured to solve the question as far as the data at command admit. He believes that should the consumption multiply for rather more than a century at its present rate, the average depth of our coal mines would be so reduced that we could not long continue our present rate of progress.* "*We have to make the momentous choice,*" *he believes,* "*between brief greatness and long-continued prosperity.*"—"*The question of our supply of coal,*" *says the* PALL MALL GAZETTE, "*becomes a question obviously of life or death.* . . . *The whole case is stated with admirable clearness and cogency.* . . . *We may regard his statements as unanswered and practically established.*"

THE THEORY OF POLITICAL ECONOMY. 8vo. 9s.

> *In this work Professor Jevons endeavours to construct a theory of Political Economy on a mathematical or quantitative basis, believing that many of the commonly received theories in this science are perniciously erroneous. The author here attempts to treat Economy as the Calculus of Pleasure and Pain, and has sketched out, almost irrespective of previous opinions, the form which the science, as it seems to him, must ultimately take. The theory consists in applying the differential calculus to the familiar notions of Wealth, Utility, Value, Demand, Supply, Capital, Interest, Labour, and all the other notions belonging to the daily operations of industry. As the complete theory of almost every other science involves the use of that calculus, so, the author thinks, we cannot have a true theory of Political Economy without its aid.* "*Professor Jevons has done*

invaluable service by courageously claiming political economy to be strictly a branch of Applied Mathematics."—WESTMINSTER REVIEW.

Martin.—THE STATESMAN'S YEAR-BOOK: A Statistical and Historical Annual of the States of the Civilized World. Handbook for Politicians and Merchants for the year 1873. By FREDERICK MARTIN. Tenth Annual Publication. Revised after Official Returns. Crown 8vo. 10s. 6d.

The Statesman's Year-Book is the only work in the English language which furnishes a clear and concise account of the actual condition of all the States of Europe, the civilized countries of America, Asia, and Africa, and the British Colonies and Dependencies in all parts of the world. The new issue of the work has been revised and corrected, on the basis of official reports received direct from the heads of the leading Governments of the world, in reply to letters sent to them by the Editor. Through the valuable assistance thus given, it has been possible to collect an amount of information, political, statistical, and commercial, of the latest date, and of unimpeachable trustworthiness, such as no publication of the same kind has ever been able to furnish. "As indispensable as Bradshaw."—TIMES.

Phillimore.—PRIVATE LAW AMONG THE ROMANS, from the Pandects. By JOHN GEORGE PHILLIMORE, Q.C. 8vo. 16s.

The author's belief that some knowledge of the Roman System of Municipal Law will contribute to improve our own, has induced him to prepare the present work. His endeavour has been to select those parts of the Digest which would best show the grand manner in which the Roman jurist dealt with his subject, as well as those which most illustrate the principles by which he was guided in establishing the great lines and propositions of jurisprudence, which every lawyer must have frequent occasion to employ. "Mr. Phillimore has done good service towards the study of jurisprudence in this country by the production of this volume. The work is one which should be in the hands of every student."—ATHENÆUM.

Smith.—Works by Professor GOLDWIN SMITH :—
A LETTER TO A WHIG MEMBER OF THE SOUTHERN INDEPENDENCE ASSOCIATION. Extra fcap. 8vo. 2s.

Smith (Prof. G.)—*continued.*

This is a Letter, written in 1864, *to a member of an Association formed in this country, the purpose of which was "to lend assistance to the Slave-owners of the Southern States in their attempt to effect a disruption of the American Commonwealth, and to establish an independent Power, having, as they declare, Slavery for its cornerstone." Mr. Smith endeavours to show that in doing so they would have committed a great folly and a still greater crime. Throughout the Letter many points of general and permanent importance are discussed.*

THREE ENGLISH STATESMEN: PYM, CROMWELL, PITT. A Course of Lectures on the Political History of England. Extra fcap. 8vo. New and Cheaper Edition. 5s.

"*A work which neither historian nor politician can safely afford to neglect.*"—SATURDAY REVIEW. "*There are outlines, clearly and boldly sketched, if mere outlines, of the three Statesmen who give the titles to his lectures, which are well deserving of study.*"—SPECTATOR.

Social Duties Considered with Reference to the ORGANIZATION OF EFFORT IN WORKS OF BENEVOLENCE AND PUBLIC UTILITY. By a MAN OF BUSINESS. (WILLIAM RATHBONE.) Fcap. 8vo. 4s. 6d.

The contents of this valuable little book are—I. "Social Disintegration." II. "Our Charities—Done and Undone." III. "Organization and Individual Benevolence—their Achievements and Shortcomings." IV. " Organization and Individualism—their Co-operation Indispensable." V. "Instances and Experiments." VI. "The Sphere of Government." "Conclusion." The views urged are no sentimental theories, but have grown out of the practical experience acquired in actual work. "Mr. Rathbone's earnest and large-hearted little book will help to generate both a larger and wiser charity."—BRITISH QUARTERLY.

Stephen (C. E.)—THE SERVICE OF THE POOR; Being an Inquiry into the Reasons for and against the Establishment of Religious Sisterhoods for Charitable Purposes. By CAROLINE EMILIA STEPHEN. Crown 8vo. 6s. 6d.

Miss Stephen defines Religious Sisterhoods as "associations, the organization of which is based upon the assumption that works of charity are either acts of worship in themselves, or means to an end, that end being the spiritual welfare of the objects or the performers of those works." Arguing from that point of view, she devotes the first part of her volume to a brief history of religious associations, taking as specimens—I. The Deaconesses of the Primitive Church. II. The Béguines. III. The Third Order of S. Francis. IV. The Sisters of Charity of S. Vincent de Paul. V. The Deaconesses of Modern Germany. In the second part, Miss Stephen attempts to show what are the real wants met by Sisterhoods, to what extent the same wants may be effectually met by the organization of corresponding institutions on a secular basis, and what are the reasons for endeavouring to do so. "*The ablest advocate of a better line of work in this direction than we have ever seen.*"—EXAMINER.

Stephen (J. F.)—A GENERAL VIEW OF THE CRIMINAL LAW OF ENGLAND. By JAMES FITZJAMES STEPHEN, M.A., Barrister-at-Law, Member of the Legislative Council of India. 8vo. 18s.

The object of this work is to give an account of the general scope, tendency, and design of an important part of our institutions, of which surely none can have a greater moral significance, or be more closely connected with broad principles of morality and politics, than those by which men rightfully, deliberately, and in cold blood, kill, enslave, and otherwise torment their fellow-creatures. The author believes it possible to explain the principles of such a system in a manner both intelligible and interesting. The Contents are—I. "The Province of the Criminal Law." II. "Historical Sketch of English Criminal Law." III. "Definition of Crime in General." IV. "Classification and Definition of Particular Crimes." V. "Criminal Procedure in General." VI. "English Criminal Procedure." VII. "The Principles of Evidence in Relation to the Criminal Law." VIII. "English Rules of Evidence." IX. "English Criminal Legislation." The last 150 *pages are occupied with the discussion of a number of important cases. "Readers feel in his book the confidence which attaches to the writings of a man who has a great practical acquaintance with the matter of which he writes, and lawyers will agree that it fully satisfies the standard of professional accuracy."* —SATURDAY REVIEW. "*His style is forcible and perspicuous, and singularly free from the unnecessary use of professional terms.*"—SPECTATOR.

STREETS AND LANES OF A CITY: being the Reminiscences of AMY DUTTON. With a Preface by the BISHOP OF SALISBURY. New and Cheaper Edition. Globe 8vo. 2s. 6d.

> This little volume records "a portion of the experience, selected out of overflowing materials, of two ladies, during several years of devoted work as district parochial visitors in a large population in the North of England." The "Reminiscences of Amy Dutton" serve to illustrate the line of argument adopted by Miss Stephen in her work on the "Service of the Poor," because they show that as in one aspect the lady visitor may be said to be a link between rich and poor, in another she helps to blend the "religious" life with the "secular," and in both does service of extreme value to the Church and Nation. "A record only too brief of some of the real portraits of humanity, painted by a pencil, tender indeed and sympathetic, but with too clear a sight, too ready a sense of humour, and too conscientious a spirit ever to exaggerate, extenuate, or aught set down in malice."—GUARDIAN.

Thornton.—ON LABOUR: Its Wrongful Claims and Rightful Dues; Its Actual Present State and Possible Future. By WILLIAM THOMAS THORNTON, Author of "A Plea for Peasant Proprietors," etc. Second Edition, revised. 8vo. 14s.

> The object of this volume is to endeavour to find "a cure for human destitution," the search after which has been the passion and the work of the author's life. The work is divided into four books, and each book into a number of chapters. Book I. "Labour's Causes of Discontent." II. "Labour and Capital in Debate." III. "Labour and Capital in Antagonism." IV. "Labour and Capital in Alliance." All the highly important problems in Social and Political Economy connected with Labour and Capital are here discussed with knowledge, vigour, and originality, and for a noble purpose. The new edition has been thoroughly revised and considerably enlarged. "We cannot fail to recognize in his work the result of independent thought, high moral aim, and generous intrepidity in a noble cause. . . . . A really valuable contribution. The number of facts accumulated, both historical and statistical, make an especially valuable portion of the work."—WESTMINSTER REVIEW.

WORKS CONNECTED WITH THE SCIENCE OR THE HISTORY OF LANGUAGE.

(*For Editions of Greek and Latin Classical Authors, Grammars, and other School works, see* EDUCATIONAL CATALOGUE.)

Abbott.—A SHAKESPERIAN GRAMMAR: An Attempt to illustrate some of the Differences between Elizabethan and Modern English. By the Rev. E. A. ABBOTT, M.A., Head Master of the City of London School. For the Use of Schools. New and Enlarged Edition. Extra fcap. 8vo. 6s.

The object of this work is to furnish students of Shakespeare and Bacon with a short systematic account of some points of difference between Elizabethan Syntax and our own. The demand for a third edition within a year of the publication of the first, has encouraged the author to endeavour to make the work somewhat more useful, and to render it, as far as possible, a complete book of reference for all difficulties of Shakesperian Syntax or Prosody. For this purpose the whole of Shakespeare has been re-read, and an attempt has been made to include within this edition the explanation of every idiomatic difficulty (where the text is not confessedly corrupt) that comes within the province of a grammar as distinct from a glossary. The great object being to make a useful book of reference for students and for classes in schools, several Plays have been indexed so fully, that with the aid of a glossary and historical notes the references will serve for a complete commentary. "A critical inquiry, conducted with great skill and knowledge, and with all the appliances of modern philology."—PALL MALL GAZETTE. *"Valuable not only as an aid to the critical study of Shakespeare, but as tending to familiarize the reader with Elizabethan English in general."*—ATHENÆUM.

Besant.—STUDIES IN EARLY FRENCH POETRY. By WALTER BESANT, M.A. Crown 8vo. 8s. 6d.

A sort of impression rests on most minds that French literature begins with the "siècle de Louis Quatorze;" any previous literature being for the most part unknown or ignored. Few know anything of the enormous literary activity that began in the thirteenth century, was carried on by Rulebeuf, Marie de France, Gaston de Foix, Thibault de Champagne, and Lorris; was fostered by Charles of Orleans, by Margaret of Valois, by Francis the First; that gave a crowd of versifiers to France, enriched, strengthened, developed, and fixed the French language, and prepared the way for Corneille and for Racine. The present work aims to afford information and direction touching these early efforts of France in poetical literature. "In one moderately sized volume he has contrived to introduce us to the very best, if not to all of the early French poets."—ATHENÆUM. "Industry, the insight of a scholar, and a genuine enthusiasm for his subject, combine to make it of very considerable value."—SPECTATOR.

Hales.—LONGER ENGLISH POEMS. With Notes, Philological and Explanatory, and an Introduction on the Teaching of English. Chiefly for use in Schools. Edited by J W. HALES, M.A., late Fellow and Assistant Tutor of Christ's College, Cambridge; Lecturer in English Literature and Classical Composition at King's College School, London; &c. &c. Extra fcap. 8vo. 4s. 6d.

This work has been in preparation for some years, and part of it has been used as a class-book by the Editor for the last two years. It is intended as an aid to the Critical study of English Literature, and contains one or more of the larger poems, each complete, of prominent English Authors from Spenser to Shelley, including Burns' Saturday Night *and* Twa Dogs. *In all cases the original spelling and the text of the best editions have been given; only in one or two poems has it been deemed necessary to make slight omissions and changes, that the "reverence due to boys might be well observed." The latter half of the volume is occupied with copious notes, critical, etymological, and explanatory, calculated to give the learner much insight in the structure and connection of the English tongue. An Index to the notes is appended.*

Helfenstein (James).—A COMPARATIVE GRAMMAR OF THE TEUTONIC LANGUAGES : Being at the same time a Historical Grammar of the English Language, and comprising Gothic, Anglo-Saxon, Early English, Modern English, Icelandic (Old Norse), Danish, Swedish, Old High German, Middle High German, Modern German, Old Saxon, Old Frisian, and Dutch. By JAMES HELFENSTEIN, Ph.D. 8vo. 18s.

This work traces the different stages of development through which the various Teutonic languages have passed, and the laws which have regulated their growth. The reader is thus enabled to study the relation which these languages bear to one another, and to the English language in particular, to which special attention is devoted throughout. In the chapters on Ancient and Middle Teutonic languages no grammatical form is omitted the knowledge of which is required for the study of ancient literature, whether Gothic or Anglo-Saxon or Early English. To each chapter is prefixed a sketch showing the relation of the Teutonic to the cognate languages, Greek, Latin, and Sanskrit. Those who have mastered the book will be in a position to proceed with intelligence to the more elaborate works of Grimm, Bopp, Pott, Schleicher, and others.

Morris.—HISTORICAL OUTLINES OF ENGLISH ACCIDENCE, comprising Chapters on the History and Development of the Language, and on Word-formation. By the Rev. RICHARD MORRIS, LL.D., Member of the Council of the Philol. Soc., Lecturer on English Language and Literature in King's College School, Editor of "Specimens of Early English," etc., etc. **Second Edition.** Fcap. 8vo. 6s.

Dr. Morris has endeavoured to write a work which can be profitably used by students and by the upper forms in our public schools. His almost unequalled knowledge of early English Literature renders him peculiarly qualified to write a work of this kind; and English Grammar, he believes, without a reference to the older forms, must appear altogether anomalous, inconsistent, and unintelligible. In the writing of this volume, moreover, he has taken advantage of the researches into our language made by all the most eminent scholars in England, America, and on the Continent. The author shows the place of English among the languages of the world, expounds clearly and with great minuteness "Grimm's Law," gives a brief history of the English language and an account of the various dialects, investigates the history and principles of Phonology, Orthography, Accent, and Etymology, and devotes several chapters to the consideration of the various Parts of Speech, and the final one to Derivation and Word-formation.

Peile (John, M.A.)—AN INTRODUCTION TO GREEK AND LATIN ETYMOLOGY. By JOHN PEILE, M.A., Fellow and Assistant Tutor of Christ's College, Cambridge, formerly Teacher of Sanskrit in the University of Cambridge. New and revised Edition. Crown 8vo. 10s. 6d.

These Philological Lectures are the result of Notes made during the author's reading for some years previous to their publication. These Notes were put into the shape of lectures, delivered at Christ's College, as one set in the "Intercollegiate" list. They have been printed with some additions and modifications, but substantially as they were delivered. "The book may be accepted as a very valuable contribution to the science of language."—SATURDAY REVIEW.

Philology.—THE JOURNAL OF SACRED AND CLASSICAL PHILOLOGY. Four Vols. 8vo. 12s. 6d.

THE JOURNAL OF PHILOLOGY. New Series. Edited by W. G. CLARK, M.A., JOHN E. B. MAYOR, M.A., and W. ALDIS WRIGHT, M.A. Nos. I., II., III., and IV. 8vo. 4s. 6d. each. (Half-yearly.)

Roby (H. J.)—A GRAMMAR OF THE LATIN LANGUAGE, FROM PLAUTUS TO SUETONIUS. By HENRY JOHN ROBY, M.A., late Fellow of St. John's College, Cambridge. Part I. containing:—Book I. Sounds. Book II. Inflexions. Book III. Word Formation. Appendices. Crown 8vo. 8s. 6d.

This work is the result of an independent and careful study of the writers of the strictly Classical period, the period embraced between the time of Plautus and that of Suetonius. The author's aim has been to give the facts of the language in as few words as possible. It will be found that the arrangement of the book and the treatment of the various divisions differ in many respects from those of previous grammars. Mr. Roby has given special prominence to the treatment of Sounds and Word-formation; and in the First Book he has done much towards settling a discussion which is at present largely engaging the attention of scholars, viz., the Pronunciation of the Classical languages. In the full Appendices will be found various valuable details still further illustrating the subjects discussed in the text. The author's reputation as a scholar and critic is already well known, and the publishers are encouraged to believe that his present work will take its place as perhaps the most original, exhaustive, and scientific grammar of the Latin language that has ever issued from the British press. "The book is marked by the clear and practical insight of a master in his art. It is a book which would do honour to any country."—ATHENÆUM. "Brings before the student in a methodical form the best results of modern philology bearing on the Latin language."—SCOTSMAN.

Taylor (Rev. Isaac).—WORDS AND PLACES; or, Etymological Illustrations of History, Ethnology, and Geography. By the Rev. ISAAC TAYLOR. New Edition, thoroughly revised and condensed for School use.   [In the press.

This work, as the SATURDAY REVIEW acknowledges, "is one which stands alone in our language." The subject is one acknowledged to be of the highest importance as a handmaid to History, Ethnology, Geography, and even to Geology; and Mr. Taylor's work has taken its place as the only English authority of value on the subject. Not only is the work of the highest value to the student, but will be found full of interest to the general reader, affording him wonderful peeps into the past life and wanderings of the restless race to which he belongs. Every assistance is given in the way of specially prepared Maps, Indexes, and Appendices; and to anyone who wishes to pursue the study of the subject further, the Bibliographical List of Books will be found invaluable. The NONCONFORMIST says, "The historical importance of ehe subject can scarcely be exaggerated." "His book," the READER says, "will be invaluable to the student of English history." "As all cultivated minds feel curiosity about local names, it may be expected that this will become a household book," says the GUARDIAN.

Trench.—Works by R. CHENEVIX TRENCH, D.D., Archbishop of Dublin. (For other Works by the same Author, see THEOLOGICAL CATALOGUE.)

D

Trench (R. C.)—*continued.*

Archbishop Trench has done much to spread an interest in the history of our English tongue. He is acknowledged to possess an uncommon power of presenting, in a clear, instructive, and interesting manner, the fruit of his own extensive research, as well as the results of the labours of other scientific and historical students of language; while, as the ATHENÆUM *says, "his sober judgment and sound sense are barriers against the misleading influence of arbitrary hypotheses."*

SYNONYMS OF THE NEW TESTAMENT. New Edition, enlarged. 8vo. cloth. 12*s*.

The study of synonyms in any language is valuable as a discipline for training the mind to close and accurate habits of thought; more especially is this the case in Greek—"a language spoken by a people of the finest and subtlest intellect; who saw distinctions where others saw none; who divided out to different words what others often were content to huddle confusedly under a common term." This work is recognized as a valuable companion to every student of the New Testament in the original. This, the Seventh Edition, has been carefully revised, and a considerable number of new synonyms added. Appended is an Index to the synonyms, and an Index to many other words alluded to or explained throughout the work. "He is," the ATHENÆUM *says, "a guide in this department of knowledge to whom his readers may entrust themselves with confidence."*

ON THE STUDY OF WORDS Lectures Addressed (originally) to the Pupils at the Diocesan Training School, Winchester. Fourteenth Edition, revised and enlarged. Fcap. 8vo. 4*s*. 6*d*.

This, it is believed, was probably the first work which drew general attention in this country to the importance and interest of the critical and historical study of English. It still retains its place as one of the most successful if not the only exponent of those aspects of Words of which it treats. The subjects of the several Lectures are—I. "Introductory." II. "On the Poetry of Words." III. "On the Morality of Words." IV. "On the History of Words." V. "On the Rise of New Words." VI. "On the Distinction of Words." VII. "The Schoolmaster's Use of Words."

Trench (R. C.)—*continued.*

ENGLISH PAST AND PRESENT. Seventh Edition, revised and improved. Fcap. 8vo. 4s. 6d.

This is a series of eight Lectures, in the first of which Archbishop Trench considers the English language as it now is, decomposes some specimens of it, and thus discovers of what elements it is compact. In the second Lecture he considers what the language might have been if the Norman Conquest had never taken place. In the following six Lectures he institutes from various points of view a comparison between the present language and the past, points out gains which it has made, losses which it has endured, and generally calls attention to some of the more important changes through which it has passed, or is at present passing.

A SELECT GLOSSARY OF ENGLISH WORDS USED FORMERLY IN SENSES DIFFERENT FROM THEIR PRESENT. Third Edition. Fcap. 8vo. 4s.

This alphabetically arranged Glossary contains many of the most important of those English words which in the course of time have gradually changed their meanings. The author's object is to point out some of these changes, to suggest how many more there may be, to show how slight and subtle, while, yet most real, these changes have often been, to trace here and there the progressive steps by which the old meaning has been put off and the new put on—the exact road which a word has travelled. The author thus hopes to render some assistance to those who regard this as a serviceable discipline in the training of their own minds or the minds of others. Although the book is in the form of a Glossary, it will be found as interesting as a series of brief well-told biographies.

ON SOME DEFICIENCIES IN OUR ENGLISH DICTIONARIES: Being the substance of Two Papers read before the Philological Society. Second Edition, revised and enlarged. 8vo. 3s.

Wood.—Works by H. T. W. WOOD, B.A., Clare College, Cambridge:—

THE RECIPROCAL INFLUENCE OF ENGLISH AND FRENCH LITERATURE IN THE EIGHTEENTH CENTURY. Crown 8vo. 2s. 6d.

CHANGES IN THE ENGLISH LANGUAGE BETWEEN THE PUBLICATION OF WICLIF'S BIBLE AND THAT OF THE AUTHORIZED VERSION ; A.D. 1400 to A.D. 1600. Crown 8vo. 2s. 6d.

> This Essay gained the Le Bas Prize for the year 1870. Besides the Introductory Section explaning the aim and scope of the Essay, there are other three Sections and three Appendices. Section II. treats of " English before Chaucer." III. " Chaucer to Caxton." IV. " From Caxton to the Authorized Version."—Appendix: I. " Table of English Literature," A.D. 1300—A.D. 1611. II. " Early English Bible." III. " Inflectional Changes of the Verb." This will be found a most valuable help in the study of our language during the period embraced in the Essay. "As we go with him," the ATHENÆUM says, "we learn something new at every step."

Yonge.—HISTORY OF CHRISTIAN NAMES. By CHARLOTTE M. YONGE, Author of "The Heir of Redclyffe." Two Vols. Crown 8vo. 1l. 1s.

> Miss Yonge's work is acknowledged to be the authority on the interesting subject of which it treats. Until she wrote on the subject, the history of names—especially Christian Names as distinguished from Surnames—had been but little examined ; nor why one should be popular and another forgotten—why one should flourish throughout Europe, another in one country alone, another around some petty district. In each case she has tried to find out whence the name came, whether it had a patron, and whether the patron took it from the myths or heroes of his own country, or from the meaning of the words. She has then tried to classify the names, as to treat them merely alphabetically would destroy all their interest and connection. They are classified first by language, beginning with Hebrew and coming down through Greek and Latin to Celtic, Teutonic, Slavonic, and other sources, ancient and modern ; then by meaning or spirit. "An almost exhaustive treatment of the subject . . . The painstaking toil of a thoughtful and cultured mind on a most interesting theme."—LONDON QUARTERLY.

R. CLAY, SONS, AND TAYLOR, PRINTERS, LONDON.

www.ingramcontent.com/pod-product-compliance
Lightning Source LLC
Chambersburg PA
CBHW020911230426
43666CB00008B/1411